The Limits of Science

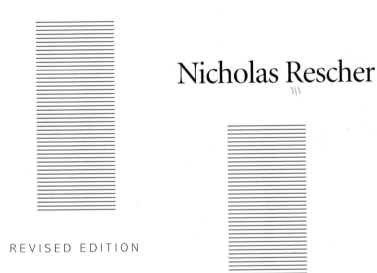

Nicholas Rescher

REVISED EDITION

UNIVERSITY OF PITTSBURGH PRESS

Published by the University of Pittsburgh Press, Pittsburgh, Pa., 15261

Copyright © 1999, University of Pittsburgh Press

Originally published 1984 by University of California Press

Copyright © 1984 by the Regents of the University of California

All rights reserved

Manufactured in the United States of America

Printed on acid-free paper

10 9 8 7 6 5 4 3 2 1

Library of Congress Cataloging-in-Publication Data
Rescher, Nicolas
   the limits of science / Nicholoas Rescher.—Rev. ed.
     p. cm.
Includes bibliographical references and indexes.
  ISBN 0-8229-5713-2 (paper : acid-free paper)
  1. Science—Philosophy. I. Title.
  Q175 .R393327 1999

  501—DC21   99-006975

A CIP catalog record for this book is available from the British Library.

For Gereon Wolters in cordial friendship

CONTENTS

Preface xi

Introduction 1

1. Question Dynamics and Problems of Scientific Completeness 5
   1. The Role of Presuppositions 5
   2. Question Dissolution 8
   3. Kant's Principle of Question Propagation 12
   4. Cognitive Completeness: Question-Answering (or "Erotetic") Completeness 15

2. Questions and Scientific Progress 19
   1. Question Dialectics and Scientific Progress 19
   2. The Lessons of History 23
   3. The Pragmatic Dimension of Progress 26

3. The Instability of Science 29
   1. The Comparative Fragility of Science: Scientific Claims as Mere Estimates 29
   2. Fallibilism and the Distinction Between Our (Putative) Truth and the Real Truth 34
   3. Cognitive Copernicanism 36
   4. The Problem of Progress 38

4. Complexity Escalation as an Obstacle to Completing Science 43
   1. Spencer's Law: The Dynamics of Cognitive Complexity 44
   2. The Principle of Least Effort and the Methodological Status of Simplicity-Preference in Science 46

3. Complexification and the Disintegration of Science   50
   4. The Expansion of Science   54
   5. The Law of Logarithmic Returns   56
   6. The Rationale and Implication of the Law   60
   7. The Growth of Knowledge   61
   8. The Centrality of Quality and Its Implications   63

5. Against Convergentism   66

   1. The Diminishing-Returns View of Scientific Progress and Its Flaws   66
   2. A Critique of the Self-Correction Thesis   68
   3. The Instability of Science: The Role of Conceptual Innovation   72
   4. The Potential Limitlessness of Scientific Change   77
   5. The Role of Cognitive Limits   81
   6. Scientific Changes Maintain a Uniform Level of Significance   85

6. Question Dynamics and Problems of Scientific Completeness   87

   1. The Impracticability of an All-Purpose Predictive Engine   87
   2. Problems of Reflexivity and Metaprediction   91

7. The Unpredictability of Future Science   94

   1. Difficulties in Predicting Future Science   94
   2. Present Science Cannot Speak for Future Science   103
   3. Against Domain Limitations   108

8. Against Insolubilia   111

   1. The Idea of Insolubilia   111
   2. The Reymond-Haeckel Controversy   113
   3. Some Purported Scientific Insolubilia   116
   4. The Infeasibility of Identifying Insolubilia   123

9. The Price of an Ultimate Theory   128

   1. The Principle of Sufficient Reason   128
   2. The Idea of an Ultimate Theory   131

3. An Aporetic Situation   134
4. A Way Out of the Impasse   135
5. Implications   140
6. Historical Postscript   142

10. The Theoretical Unrealizability of Perfected Science   145

1. Conditions of Perfected Science   145
2. Theoretical Adequacy: Issues of Erotetic Completeness   147
3. Pragmatic Completeness   151
4. Predictive Completeness   153
5. Temporal Finality   155
6. The Dispensability of Perfection   157
7. "Perfected Science" as an Idealization that Affords a Useful Contrast Conception   159
8. Science and Reality   161

11. The Practical Infeasibility of Perfecting Science   166

1. Technological Escalation   166
2. Rising Costs   170
3. Economic Requirements Spell Economic Limitations   172

12. Can Computers Overcome Our Limitations?   177

1. Could Computers Overcome Our Limitations?   177
2. General-Principle Limits Are not Meaningful Limitations   179
3. Practical Limits: Inadequate Information   181
4. Practical Limits: Transcomputability and Real-Time Processing Difficulties   182
5. Practical Limits: Limitations of Representation in Matters of Detail Management   183
6. Performative Limits of Prediction-Self-Insight Obstacles   184
7. Performative Limits: A Deeper Look   185
8. Contrast with Algorithmic Decision Theory   187
9. A Computer Insolubilium   188

10. The Human Element: Can People Solve Problems that Computers Cannot? 189

11. Potential Difficulties 191

Appendix to Chapter 12: On the Plausibility of T_1 and T_2 193

13. Extraterrestrial Science (Could Aliens Overcome Our Limitations?) 197

1. Could Science in Another Setting Overcome the Limitations of Our Human Science? 197
2. The Potential Diversity of "Science" 199
3. The One-World, One-Science Argument 204
4. Comparability and Judgments of Relative Advancement 208
5. First Principles 211
6. The Implausibility of Being Outdistanced 216

Appendix: References for Chapter 13 219

14. The Limits of Quantification in Human Affairs 223

1. The Problem 223
2. Quantification Versus Measurement: What Makes a Number Meaningful? 225
3. Problematic Measurements 232
4. Quality of Life as an Example 233
5. Fallacies of Quantification 234
6. Larger Vistas 238

15. The Limited Province of Natural Science 241

1. Knowledge as One Good Among Others 242
2. Scientific Knowledge as One Mode of Knowledge 243
3. The Autonomy of Science 248
4. Conclusion 250

Notes 253

Index 279

PREFACE

This is the outgrowth of a long-standing preoccupation with the theory of scientific inquiry, which has issued in such earlier books as *Scientific Explanation* (New York, 1970), *Scientific Progress* (Oxford, 1978), *Cognitive Systematization* (Oxford, 1979), and *Empirical Inquiry* (Totowa, N.J., 1981). In particular, it stands in apposition to *Scientific Progress*. That work focused on practical, and specifically economic, obstacles to scientific progress. This book, in contrast, largely abstracts from such practical obstacles to focus on the existence of limits or limitations on scientific inquiry that could in principle preclude the full realization of the aims of science. Its deliberations are accordingly less empirical and more philosophical, addressing themselves primarily to those theoretical general principles that characterize the very nature of the scientific enterprise.

During the Trinity Term of 1982, I presented a part of this material in a set of lectures given in the Faculty of Literae Humaniores of the University of Oxford at the kind invitation of the subfaculty of philosophy. I want to express my appreciation to Corpus Christi College for providing me with an academic home away from home on this as on many previous occasions.

I would like to thank Timo Airaksinen, Bryson Brown, and Geoffrey Sayre-McCord, who read the material in draft form and made helpful suggestions for its improvement. And I am grateful to Linda Butera and Donna Williams, who have struggled to turn my hen scratches into a usable typescript, persevering undaunted through a seemingly endless sequence of revisions. I am also grateful to Christina Romanelli for her help.

This new edition of the book owes its existence to the interest of some Pittsburgh colleagues and the supportive response of Cynthia Miller, Director of our University Press. I am grateful for their initiative. And I owe many thanks to Estelle Burris for her patient and competent help in preparing the material for publication.

Apart from the removal of some infelicities, the present version of the book differs from its predecessor as follows. Of the twelve chapters of that book, two have dropped away (chapter 1 on "A Mistaken View of the Completeness of Science," and chapter 4, "The Potential Limitlessness of Science"—much of the latter being absorbed into chapter 5 of the present work). On the other side, five new chapters have been added, specifically numbers 4, 6, 9, 12, and 14. The result is a book that is not only substantially larger but also (so I hope) substantially stronger.

Nicholas Rescher
Pittsburgh, Pennsylvania
June 1999

The Limits of Science

Introduction

It is important—and more so nowadays than ever—to have a realistic appreciation of just what science can and cannot be expected to accomplish, for the strident calls of ideological extremes surround us. On one side lies the exaggerated scientism of the "science can do anything" persuasion that sees science in larger-than-life terms as an all-powerful be-all and end-all. On the other side lies the antiscientism or even irrationalism that sees science as a dangerous luxury, a costly diversion we would be well-advised to abandon altogether. And these positions are closely related: if we have high (or low) expectations, it is easy to be propelled from the one extreme to the other. The sensible middle path between unrealistic extremes lies in a just appreciation of the powers and limitations of science to avoid the disappointments of unrealistically large expectations or the debility of unrealistically small ones.

Natural science has become an enormously complex and costly venture, to which any advanced society nowadays quite properly dedicates massive resources. It is highly important that people should have a realistic understanding of the inherent limits of the scientific enterprise so as to prevent inflated and unrealistic expectations, and thus to avoid the backlash of reproach, recrimination, and alienation to which the disappointment of such unreasonable expectations could all too easily lead. The present book is an attempt to offer some contributions toward achieving such a just appreciation.

In deliberating about the limits of science, three questions must be distinguished:

1. How far *might* science actually go: what are the *practical* limits on science?
2. How far *should* science go: what are the prudential and the *moral* limits on science?

3. How far *could* science go in principle: what are the *theoretical* limits on science?

This book concentrates mainly on the last of these issues. Accordingly, the book does not deal with matters of ideology. It concerns itself with limits relating to the internal workings of the scientific enterprise and to the capacity of science to "do the job" that defines its characteristic mission. It considers what science *can* discover, and not what it *should* discover. Whether there are certain matters that science had best leave alone in the interests of individual comfort, social harmony, or public safety is a question with which these pages do not deal at all.[1] Again, the book does not address the complaints of those who maintain that the consequences of man's cultivation of science and technology are on balance negative and that their pursuit diminishes the quality of human life and impedes the realization of human values.[2] The focal question for these deliberations is how far natural science can go toward realizing its mission, and not whether pursuit of this mission may somehow prove to be detrimental to other human interests.

Moreover, the reader who expects to find a status report on present-day science—an overview of its achievements and a survey of various unsolved problems that remain on its agenda—will be gravely disappointed in the book. Its concern is wholly with those limits and limitations that characterize science in consequence of general principles, and its task is not a descriptive characterization of the present condition of science but a philosophical exploration of the fundamentals affecting the enterprise regardless of time or place. To address the contingent realities of the present state of science would be to cut ourselves off from those deeper, pervasive, and timeless issues that are the focus of our present concerns.

A modicum of terminological regimentation will facilitate an overview of the present deliberations. With respect to a given task or accomplishment, one can ask: does it belong to the mission of natural science, or not—falling outside science into some other domain (aesthetics or history or whatever)? If the answer to this question is affirmative—if the task at issue does indeed belong to the mission or mandate of science—we can proceed to ask whether science is, in principle, able to accomplish it. And if the answer to this question is negative, then science clearly encounters a limit at this stage. This sequence of questions provides distinctions that clarify the purpose of our discussion: *does the task at issue actually belong to the proper domain of science?* (See figure I.1.).

Such a distinction between *disabilities* (domain-external issues, limits, incapacities) on the one hand, and actual *deficiencies* on the other, is essential to the realization of clarity in the problem context of our discussion. It facilitates placing a proper emphasis on the fact that our present concern is with the lim-

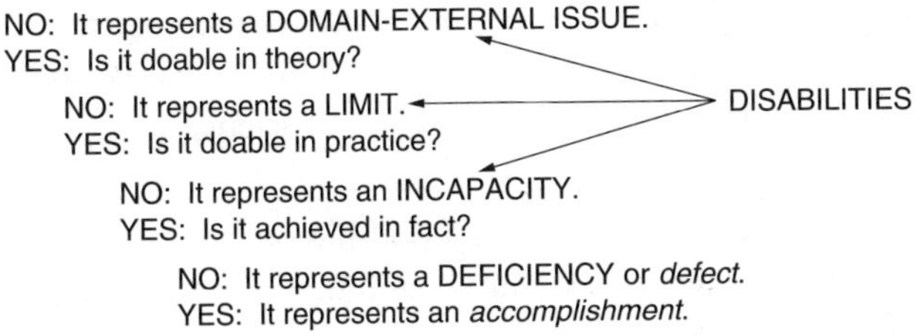

```
NO: It represents a DOMAIN-EXTERNAL ISSUE.
YES: Is it doable in theory?
    NO: It represents a LIMIT. ──────────▶ DISABILITIES
    YES: Is it doable in practice?
        NO: It represents an INCAPACITY.
        YES: Is it achieved in fact?
            NO: It represents a DEFICIENCY or defect.
            YES: It represents an accomplishment.
```

Figure I.1

its and incapacities of natural science, and not with its deficiencies. Our deliberations are concerned not with what science *has not* achieved but with what it *cannot* achieve.

Both limits and incapacities reflect, as it were, *domain-internal disabilities* of natural science. They indicate the existence of issues within the mission of science that, for reasons of fundamental fact—be they theoretical (limits) or practical (incapacities)—science is simply unable to resolve. Insofar as such disabilities exist, our scientific knowledge will, always and inevitably, fall short of realizing in full the characterizing aims of the enterprise.

However, *domain-external disabilities* are something else. They revolve around the very different questions of whether these are issues that we humans would want to have resolved that could not be resolved in terms of scientific information about the world even if the scientific project were carried through to completion and perfection. Such issues, in short, would by their very nature lie outside the domain of science. They have nothing to do with defects or deficiencies, relating not to any interior shortcomings but to the mere existence of an outside.

This said, we may foreshadow that the present deliberations will lead to four significant, and significantly different, conclusions:

1. Natural science has no limits (in the presently operative, strict sense of that term). There is no reason to think, on the basis of general principles, that any issues within the domain of natural science lie beyond its capabilities (chapters 6 and 7).

2. However, there is good theoretical reason to think that natural science can never be completed: it can never manage to resolve all of its questions. Natural science thus has domain-internal incapacities (chapters 2–4).

3. There are also powerful limitations to science of practical (and ultimately economic) provenience. One cannot, however, adduce any concrete examples of irresolvable insolubilia that represent incapacities of science (chapter 8 and 9).

4. There is no question that natural science is subject to domain-external incapacities. We must recognize that various important evaluative and cognitive issues lie altogether outside the province of science as we know it (chapters 10 and 11).

These four points afford a preview of the main findings of the book. In the end, our deliberations indicate that while science does indeed have disabilities, they are nothing like those that people who talk about the limits of science usually take them to be—namely, insolubilia and capacity limitations. Science, like any other human enterprise, is inevitably limited, but there is no reason whatsoever to think that it must reach a dead end. The sorts of disabilities to which science is indeed subject—fallibilism, instability, and thus inability to arrive at anything ultimate and definitive—are simply the reverse side of its strength as an endlessly versatile intellectual instrument capable of accommodating itself to ever-changing cognitive circumstances.

1 Question Dynamics and Problems of Scientific Completeness

Synopsis
(1) Questions always have presuppositions that embed them within a preexisting state of knowledge—scientific questions preeminently so. (2) Accordingly, questions not only will arise from new knowledge (when new presuppositions become available) but will also be dissolved by new knowledge (when old presuppositions are abandoned). Change in regard to questions is every bit as important and dramatic as change in regard to knowledge. (3) In particular, we must reckon with Kant's Principle of Question Propagation, to the effect that the answers to our scientific questions always spawn further questions. (4) Science would achieve a condition of "erotetic completeness" if it attained a state in which all of its questions become answerable. But Kant's Principle indicates that such a condition cannot be realized.

1. The Role of Presuppositions

The aim of scientific inquiry is to resolve our questions about the hows and whys of natural phenomena. Now, scientific inquiry—like inquiry in general—proceeds through a process of dialectical interaction between questions and answers. To understand the workings of this process, it is necessary to consider some fundamental issues in the theory of questions, beginning with what is perhaps the most fundamental conception of this domain, that of presupposition.

A *presupposition* of a question is a thesis (or proposition) that is inherent in (and thus entailed by) each of its possible fully explicit answers.[1] For example, "What is the melting point of lead?" has innumerably many fully explicit answers, all taking the general form "m °C is the melting point of lead." All of these imply that lead indeed has a (fixed and stable) melting point. This thesis accordingly emerges as a presupposition of the question. Again, the explanatory question "How does the moon cause eclipses of the sun?" takes answers of the form "The moon causes eclipses of the sun X-wise" and therefore presupposes that the moon does indeed on occasion operate so as to produce solar eclipses. Its presuppositions inhere in the very way in which a question is formulated: regardless of which answer we endorse, the presuppositions are something we will stand committed to in any case.

The presuppositions of our questions reflect their *precommitments*: they constitute the formative background that we bring to the very posing of questions, rather than merely being something we take away as a result of answering them. The propriety of a question is accordingly predicated on the availability of its presuppositions: a question whose presuppositions are not satisfied simply does not arise. If Smith has no wife, questions about his ceasing to beat her are inappropriate. Similarly, if there were no such thing as a fixed velocity of light *in vacuo* (if this were something variable and circumstance-dependent), then it would make no sense to ask "Can anything move faster than the velocity of light?"

If one state of the art in science S includes the proposition p among its commitments, while another, S', fails to include p, then we can ask the rationale-demanding question, "Why is it that p is the case with respect to S but not with respect to S', where this presupposition fails?" When we change our mind about the correct answers to questions, we are also thereby at once involved in a change of mind as to the sorts of questions that can sensibly be asked. Discordant bodies of (putative) knowledge engender distinct bodies of questions because they provide the materials for distinct sets of background presuppositions.

A "state of scientific knowledge" S—or, more accurately, of *purported* knowledge—is always correlative with a body of questions $Q(S)$ that can be posed on its basis, that is, questions whose presuppositions it is able to assure. Thus $Q(S)$ represents the question agenda of S. A question Q belongs to it if all of its presuppositions are forthcoming from S, that is, if the question is appropriate relative to S^2. The questions of $Q(S)$ are appropriately formulable relative to S-available concepts and S-available theses; each of these questions is such that its concepts are meaningful relative to S, and all of its other presuppositions are S-true.

To pose or otherwise endorse a question is to undertake an at least tacit commitment to all of its presuppositions. And where different presuppositions are available, different bodies of questions can be raised. Accordingly, the state of the art regarding questions is inseparably geared to the state of the art regarding knowledge. When the membership of a body of accepted knowledge S

changes, so of course does that of its erotetic agenda, $Q(S)$, the corpus of questions that can appropriately be asked on its basis.

Epistemic change over the course of time accordingly relates not only to what is *known* but also to what can be *asked*. The accession of new knowledge opens up new questions. When the epistemic status of a presupposition changes from acceptance to abandonment or rejection, we have not so much the opening of new questions as the disappearance of various old ones through dissolution. Questions regarding the modus operandi of Phlogiston, the behavior of caloric fluid, the structure of the luminiferous ether, and the character of faster-than-light transmissions are all questions that have become lost to modern science because they involve presuppositions that have been abandoned. The phenomenon of presupposition reflects the fact that our questions are intimately tied to the state of the art in point of knowledge—the *putative* knowledge of the day.

Questions are always projected relative to an existing body of putative knowledge. One is not entitled to speak unqualifiedly of either questions or answers as such but only as they are seen at a certain stage of the game—by our own contemporaries, for example, or by the physicists of Kelvin's day. It is an illicit hypostatization to speak simply of "the body of scientific knowledge" without adding something like "of the late nineteenth century" or "of the present day." In the same way, it makes no sense to speak unqualifiedly of "the body of scientific questions"—no well-defined set of items is at issue here, in the absence of a particularizing delimitation. (Our symbol 'Q', is not a variable with a particular range, but a notational device.) We must accordingly be very careful about our all-statements in this context and make sure of category-limitations when embarking on declarations beginning with the locution "All scientific questions. . . ." And thus caution is particularly germane when we realize that the putative scientific knowledge of different eras can contain mutually incompatible contentions, so that their questions can be mutually preemptive in having actually incompatible presuppositions.

The idea of a question agenda accordingly cries out for temporalization. It is profitable to consider more closely the temporalized "state of questioning" at the time t, with $Q(S_t)$ as the set of questions or problems appropriately posable relative to the historical "state of knowledge" S_t that obtains at t—the problem-field as it stands at the historical juncture t, comprising all the questions explicit or implicit in the commitments of the scientific community of the time.

Membership in $Q(S_t)$ requires that the question at issue be a feasible one relative to S_t, but not necessarily that it be actually puzzled over and investigated by the concurrent practitioners of the discipline. Which questions are actually asked is something contingent and often fortuitous. Often, indeed, science acknowledges a question only after it has been answered. Only after Rayleigh worked out his theory of atmospheric dispersion did the question of the blue color of the sky acquire much significance in optical theory. Only after Darwin's theory of sexual selection did the long-recognized variation of animal mating behavior acquire importance. In science as in life, there is much to be said for the wisdom of hindsight. In any case, the question set $Q(S_t)$ represents the *potential erotetic agenda* of a body of scientific knowledge espoused at t, the entire manifold of admissible questions spawned by this body of knowledge, including alike those it can and those it cannot answer.

2. Question Dissolution

No meaningful question—no question consonant with the proprieties of language—can be improper by its very nature. Questions become illegitimate only in context, relative to what we accept or fail to accept—that is, only through the unavailability of their presuppositions.

One way in which a body of knowledge S can deal with a question is, of course, by *answering* it. Yet another, importantly different, way in which S can deal with a question is by disallowing it. S *disallows W* when there is some presupposition of Q that S does not countenance:[3] given S, we are simply not in a position to raise Q. The circumstance that S fails to fall within $Q(S)$ occurs when it has a presupposition p that is unavailable for any one of three reasons: (1) p is false relative to S (not-$p \in S$); (2) p is *undecidable* relative to S, as per claims about the mountains on the far side of the moon in the nineteenth century; or (3) p is inexpressible within S, as happens with claims about the workings of the Galenic humors in modern biochemistry. We thus arrive at three corresponding modes of question-disallowing:

1. *Impropriety*, which arises when a question has an S-relatively false presupposition. (In this case K may be said to *block* the question at issue.)

2. *Problematicity*, which arises when a question is indeterminate relative to S and thus "presumes too much," involving an indeterminate presupposition—one whose truth is simply not available.

3. *Ineffability*, which arises when a question has a presupposition that is conceptually inaccessible (that is, "meaningless") relative to S—one for whose concepts S simply fails to provide.

A question that is disallowed by a body of scientific knowledge (in any one of these three ways) may be said to be *illegitimate* with respect to it.

Consider some examples. Present-day physics rules out as improper questions predicated on the realization of transluminar velocities or on the existence of a *perpetuum mobile*.[4] Indeed, every purported law of nature rules out certain things as impossible; the acceptance of any generalization as a law of nature will block those questions based on conflicting presuppositions.

Problematicity is something else again. Given the present state of our knowledge, questions about the communicative procedures of extraterrestrial inhabitants of our galaxy will be problematic: they are premature and "presume too far." Not just their answer but their very possibility is a moot issue. Yet they are certainly not improper—they are perfectly consonant with everything that we do know. Such questions can only be posed in the hypothetical and not in the categorical mode. ("If there are extraterrestrials, how might they communicate?")

Ineffable questions are not just unanswerable but actually *unaskable* because they cannot even be posed in the given state of knowledge. We are not in a position even to formulate such a question—it lies wholly beyond the reach of the prevailing cognitive state of the art. Such questions can only be instanced through historical examples. Newton could not have wondered whether plutonium is radioactive. It is not just that he did not know what the correct answer to the question happens to be—the very question not only *did* not but actually *could* not have occurred to him, because the cognitive framework of the then-existing state of knowledge did not afford the conceptual instruments with which alone this question can be posed. Conceptual innovation involves the formulation of issues that could not even be contemplated at an earlier juncture of the cognitive state of the art—and may well abandon previously operative conceptions (for example, the luminiferous ether).

The history of science is replete with cases of this sort. In the main, today's scientific problems could not even have been contemplated a generation or two ago: their presuppositions were cognitively unavailable. Every cognitive state of the art has its characteristic conceptions that set correlative limits precluding certain matters from even arising, thereby rendering certain alien issues undiscussable. (Modern chemistry simply has no place for "affinity," nor modern physics for *vis viva*.)

It is convenient to adopt the convention that a body of knowledge S is said to *resolve* a question when S either *answers* or *disallows* it. For our factual knowledge is perfectly entitled to establish certain sorts of questions as improper—as "just not arising at all." Thus, when a certain form of motion (be it Aristotle's circles or Galileo's straight lines) is characterized as "natural" in a physical theory, then we are thereby precluded from asking why—in the absence of imposed forces—objects move in this particular manner. Or, again, considering that the half-life of a certain species of *californium* is 235 years, we must not ask—given modern quantum theory—just why a certain particular atom of this substance decayed after only one hundred hours. Such questions have presuppositions that are at odds with commitments of the body of knowledge at issue.

The question-answering capacity of a body of knowledge should accordingly be assessed in terms not of all imaginable questions but of all the proper or legitimate questions, where the cognitive corpus at issue is itself entitled to play a part in the determination of such legitimacy. When a given body of knowledge actually disallows certain questions, we cannot automatically regard its failure to provide answers for them as counting to its discredit. And in the light of such considerations, we cannot maintain that science can explain everything: the very most we can possibly claim is that "science can explain everything *explicable*; it can answer all legitimate explanatory questions"—where science is itself the controlling determiner of legitimacy.

The prospect of relative illegitimacy shows that the body of knowledge of the day not only delimits the assertions we can maintain, it also limits the range of questions we can appropriately raise. Immanuel Kant's *Critique of Pure Reason* is dedicated to the proposition that certain issues (that is to say, those of traditional metaphysics) cannot be legitimately posed at all because they are *absolutely* illegitimate—because they overstep the limits of possible experience.[5] The present deliberations pose the more mundane yet interesting prospect that certain questions are *circumstantially illegitimate*, because they transcend the limits of actual experience, in that certain presuppositions of these questions run afoul of the body of knowledge on hand.

Ignorance—the inability to resolve questions—falls into two very different types, in line with the distinction between merely unanswerable questions and actually unaskable ones. Ignorance of the finite type arises when we can grasp a question but, under the prevailing circumstances, are unable to answer it. (Think of the status in Darwin's day of questions regarding the mechanism of heredity.) Ignorance prevails at the deeper level of actual inaccessibility when we would not even pose the question—and indeed could not even understand an answer to it should one be vouchsafed us by a benevolent oracle—because presuppositions of undetermined truth-status are involved in such a way that the whole issue lies beyond the conceptual horizons of the day. It is not difficult to envisage present-day questions that exhibit surface ignorance; any practicing scientist can readily give examples of this kind from the domain of his own research. However, inaccessibility-ignorance cannot be illustrated, except in historical retrospect. Nevertheless, the fact that current ideas went unrealized at all earlier historical stages is readily amplified to the speculative prospect that some intrinsically feasible ideas may go unrealized at all historic stages whatsoever. For it is perfectly conceivable that some facts will never be recognized as such, and consequently that some questions—namely those that presuppose such theses—can never be appropriately posed.

Consider the following *Thesis of the Conservation of Questions*: "Once posable, a question always remains so. A question that can be raised appropriately in one state of the art in science will ever after continue to figure on the agenda."[6] This thesis is patently false. A presupposition of Q that is available at t relative to the state of science at that time, S_t may fail to be available at t' relative to $S_{t'}$. Questions can be "lost" by being forgotten, but they can also be dissolved in the course of scientific evolution, when the scientific community comes to abandon their presuppositions.

The coming to be and passing away of questions is a phenomenon that can be mooted on this basis. A question *arises* at the time t if it then can meaningfully be posed because all its presuppositions are then taken to be true. And a question *dissolves* at t if one or another of its previously accepted presuppositions is no longer accepted. Any state of science will remove certain questions from the agenda and dismiss them as inappropriate. Newtonian dynamics dismissed the question "What cause is operative to keep a body in movement (with a uniform velocity in a straight line) once an impressed force has set it into motion?" Modern quantum theory does not allow us to ask "What caused this atom of *californium* to disintegrate after exactly 32.53 days, rather than, say, a day or two later?" Scientific questions should thus be regarded as arising in a historical setting. They arise at some junctures and not at others; they can be born and then die away.

A change of mind about the appropriate answer to some question will unravel the entire fabric of questions that presupposed this earlier answer. For if we change our mind regarding the correct answer to one member of a chain of questions, then the whole of a subsequent course of questioning may well collapse. If we abandon the luminiferous ether as a vehicle for electromagnetic radiation, then we lose at one stroke the whole host of questions about its composition, structure, mode of operation, origin, and so on. The course of erotetic change is no less dramatic than that of cognitive change, change in the domain of questions no less than in the domain of answers.

A body of knowledge may well answer a question only provisionally, in a tone of voice so tentative or indecisive as to suggest that further information is actually needed to settle the matter with confidence. But even if it does confidently and unqualifiedly support a certain resolution, this circumstance can never be viewed as absolutely final. What is seen as the correct answer to a question at one stage of the cognitive venture may, of course, cease to be so regarded at another.[7] Given an S-relatively appropriate answer to Q we can never preclude the prospect that some better S' will eventually come about as successor to S, and that it will then transpire that some other answer—one that is actually inconsistent with the preceding one—comes to be seen as the correct answer to Q relative to S'.

To be sure, this tentativity is abstract, and, in a way, irrelevant and immaterial. In scientific inquiry, as in all other human affairs, we have to proceed from where we are. Our confidence in the answers forthcoming relative to S_n—that is, S_t with $t = n$ for *now*, the state of the cognitive art that is currently at hand—should never be impaired by doubts that represent no more than a matter of "general principles."[8] The membership of S_n would not represent our best efforts if we did not see ourselves as committed to them: they would not represent *our* truth if we did not seriously regard them as representing our best efforts to arrive at *the* truth. And so, if S_n itself gives sufficiently powerful indications on p's behalf—so powerful as to indicate the ill-advisedness of expending further resources of time and energy in pursuing the issue—then we are quite entitled to let the matter rest there for the present. All the same, we must recognize that this is an essentially practical rather than theoretical position. The economic aspect comes to the fore here: where the prospect of error is sufficiently remote, there presumably is no practical point in expending resources in an endeavor to accommodate purely hypothetical worries.[9]

3. Kant's Principle of Question Propagation

W. Stanley Jevons wrote, some hundred years ago, that "since the time of Newton and Leibniz, worlds of problems have been solved which before were hardly conceived as matters of inquiry."[10] Cognitive progress is commonly thought of in terms of the discovery of new facts—novel information about things. The situation is in fact more complicated, because not only *knowledge* but also *questions* must come into consideration. Progress on the side of questions is also cognitive progress, correlative with—and every bit as important as—progress on the side of information. The questions opened up for our consideration are as definitive a facet of a state of knowledge as are the theses that it endorses.

New knowledge can bear differently on the matter of questions. Specifically, we can discover:

1. New (that is, different) answers to old questions;

2. New questions;

3. The impropriety or illegitimacy of our old questions.

With (1) we discover that the wrong answer has been given to an old question: We uncover an error of commission in our previous question-answering endeavors. With (2) we discover that there are certain questions that have not heretofore been posed at all: we uncover an error of omission in our former question-asking endeavors. Finally, with (3) we find that one has asked the wrong question altogether: we uncover an error of commission in our former question-asking endeavors, which are now seen to rest on incorrect presuppositions (and are thus generally bound up with type [1] discoveries). Three rather different sorts of cognitive progress are thus involved here—different from one another and from the traditional view of cognitive progress in terms of an accretion of further knowledge.

The second of these modes of question-directed discovery is particularly significant. The phenomenon of the new questions was first emphasized by Immanuel Kant, who saw the development of natural science in terms of a continually evolving cycle of questions and answers, where "*every answer given on principle of experience begets a fresh question, which likewise requires its answer and thereby clearly shows the insufficiency of all scientific modes of explanation to satisfy reason.*"[11] Kant's claim suggests the following Principle of Question Propagation (Kant's Principle): "The answering of our factual (scientific) questions always paves the way to further yet unanswered questions."

How can this principle possibly be established? What is at issue here is not, of course, simply the relatively trivial point that whenever we introduce a new claim p into the family of what we accept, we can inquire into such matters as the reasons for p's being the case and the relationship of p to other facts that we accept. Rather, the issue turns on the more interesting theoretical matter that new answers change the presuppositions available for new questions. As we deepen our understanding of the world, new problem areas and new issues are bound to come to the fore; as we discover, for example, that atoms are not really "atomic" but actually have an internal composition and complexity of structure, questions about this whole "subatomic" domain become available for investigation. At bottom, Kant's Principle rests on the insight that no matter what answers are in hand, we can proceed to dig deeper by raising yet further questions about these answers themselves. Whenever we obtain new and different answers, interest is at once deflected to the issues they pose. The physicist postulates a new phenomenon—we want to know its character and modus operandi. The chemist synthesizes a new substance—we want to know how it interacts with the old ones.

The answer to a question inevitably sets the stage for yet further questions by providing new materials from which they can be drawn. This leads to a cyclic process with the following structure:

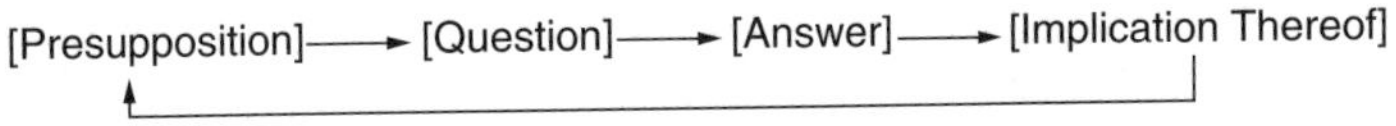

[Presupposition]⟶ [Question]⟶ [Answer]⟶ [Implication Thereof]

Figure 1.1

This cycle determines a course of inquiry set by an initial, controlling question together with the ancillary questions to which it gives rise and whose solutions are seen as facilitating its resolution. One question emerges from another in such a course of inquiry that it is not until we have answered the latter that the former becomes posable. The unfolding of such a series provides a direction of search—of research—in question-answering inquiry.

This principle of question propagation in empirical inquiry indicates a fact of importance for the theory of scientific progress. One need not claim longevity—let alone immortality—for any of the current problems to assure that there will indeed be problems ten or one hundred generations hence. (As immortal individuals are not needed to assure the immortality of the race, so immortal problems are not needed to assure the immortality of problems.)

Note, however, that Kant's Principle can be construed in two rather different ways:

i. A *universalized* mode: EACH specific (particular) question Q that can be raised on a basis of S engenders a (Q-correlative) line of questioning that leads ultimately to a question Q' whose answer lies outside of S, a question that forces an eventual shift from S to some suitably augmented or revised modification thereof.

ii. A *particularized* mode that arises when the capitalized EACH of the preceding formula is replaced by SOME.

On the first construction, science is an essentially divergent process, with questions leading to more questions in such a way that the erotetic agenda of successive stages of science is ever-increasing. This view was endorsed by W. Stanley Jevons, who wrote: "As it appears to me, the supply of new and unexplained facts is divergent in extent, so that the more we have explained, the more there is to explain."[12] The second construction is, however, a far more modest proposition, which sees science as having a question agenda of nontrivial size at every successive stage, but not necessarily a growing one, since questions may well die off by dissolution at a rate roughly equal to that of the birth of new questions.

Kant undoubtedly intended the principle in the first (universalized) sense. It is clearly more plausible and realistic to adopt it in the second, more modest (particularized) sense, which yields a thesis amply supported by historical experience: that every state-of-the-art condition of questioning $Q(S)$ ultimately yields, somewhere along the road, a line of questioning that engenders the transition from S to a different S'. The states of science are unstable: the natural course of inquiry provides an impetus by which a given state is ultimately led to give way to its successor.

The motive force of inquiry is the existence of questions that are posable relative to the body of knowledge of the day but not answerable within it. Inquiry sets afoot a sequential process, of the cyclic form depicted in figure 1.2 in which the body of scientific knowledge S and the correlative body of scientific

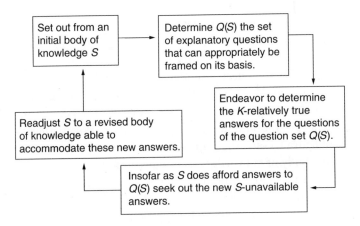

Figure 1.2 The Erotetic Dynamics of Inquiry

questions $Q(S)$ undergo continual alteration. This process gives rise to successive stages of knowledge, S_t, $S_{t'}$, $S_{t''}$, and so on, together with their associated state-of-the-art stages with regard to questions, $Q(S_t)$, $Q(S_{t'})$, $Q(S_{t''})$, and so forth.[13]

4. Cognitive Completeness: Question-Answering (or "Erotetic") Completeness

Heed of the relationship between questions and answers suggests a characteristic sense of *completeness* for bodies of knowledge, since one prime version of the idea of cognitive completeness is that "every question is answered." Every *admissible* question, that is, because the idea of an all-inclusively complete state of knowledge in which literally *all* questions are answered simply makes no sense. No matter what such a state of knowledge might be, there will, of course, be questions whose presuppositions are at variance with it, questions it does not even allow to arise—as relativity theory prohibits questions about causal processes that propagate faster than the speed of light. As we have seen, the very idea of "all questions" is inherently problematic in its literally universal construction.[14] The best we can do is to contemplate the S-relative range of "all appropriate questions." The completeness of a body of knowledge should accordingly be assessed not in terms of all imaginable questions but of all legitimate questions, where the cognitive framework at issue is itself entitled to play a part in a determination of such legitimacy.

Let us say that S_t, the state of scientific knowledge at the time t, has achieved a condition of Q-completeness (question-answering or "erotetic" completeness) if it can furnish answers to all explanatory questions throughout $Q(S_t)$, the whole range of S_t-appropriate questions. To say this is to say that every then-askable question has a then-available answer, $Q(S_t) = Q^*(S_t)$, where $Q^*(S_t)$ represents the set of all questions of $Q(S_t)$ that S_t is able to answer. This conception of Q-completeness reflects the idea of an equilibrium between questions and answers, which subsists when the questions that can be raised on the basis of a body of knowledge can be answered with recourse to this same body of knowledge (as per the situation of figure 1.2, whose flow pattern is identical with that of figure 1.1 once the amplification and revision stages of the latter are deleted). If a body of knowledge reaches this erotetic equilibrium of Q-completeness, there will no longer be any apparent need for its further development: every question in sight is resolved. But this, of course, is just what is ruled out by Kant's Principle of Question Propagation.

This circumstance is regrettable because the general idea of the question-answering (or *erotetic*) completeness of a state of knowledge is in one way a particularly attractive conception. For it does not construe completeness in terms of some external, absolutistic standard of perfect information and moot a comparison between our knowledge and that of some hypothetical, cognitively infallible being. Rather, it develops a standard of completeness that is *internal* to our cognitive horizons, approaching the matter from the standpoint of whether all the questions that *we* do (or meaningfully can) pose from where we stand on the cognitive scheme of things are questions that *we* do (or can) answer. Such completeness appraises a state of knowledge on its own terms, and not with a view to some transcendent absolute.

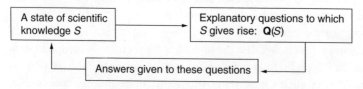

Figure 1.3 The Question-Answer Cycle

The fact remains, however, that even erotetic completeness remains elusive. As William James put it with his characteristic flair: "It seems *a priori* improbable that the truth should be so nicely adjusted to our needs and powers.... In the great boarding-house of nature, the cakes and the butter and the syrup seldom come out so even and leave the plates so clean."[15] It is accordingly of some consolation to realize that the erotetic completeness of a state of knowledge S need not necessarily betoken its comprehensiveness or sufficiency but might simply reflect the paucity of the range of explanatory questions $Q(S)$ that it is in a position to contemplate. "Explanatory completeness" is a matter of explaining everything that is deemed to need explaining. The fact that various issues can stand outside the purview of a given state of science makes its capabilities within its range distinctly less impressive. When our body of scientific knowledge S is sufficiently restricted, then its Q-completeness will merely reflect this restrictedness rather than its intrinsic adequacy.

Erotetic completeness is completeness of a narrow sort: our corpus of scientific knowledge could be erotetically complete and yet fundamentally inadequate because too myopic. Conceivably, even if improbably, a body of scientific knowledge might achieve a purely fortuitous equilibrium between problems and solutions. It could eventually be "completed" in the erotetic sense—in providing an answer to every explanatory question one can ask in the then-existing (still imperfect) state of knowledge—without our thereby being all that well off.

Suppose we could play Twenty Questions with God—certainly more than natural science can ever accomplish. The rules of the game are simple: we are allowed to ask any yes-or-no question, and God's voice booms from the heavens, indicating which answer is closer to the real truth. Now we ancient Greeks, say, ask: "Are there elements?"—*Yes*. "Are these elements the standard ones—that is to say, Earth, Air, Fire, and Water?"—*No*. Where are we to go from there? We could move down the list of substances we can think of: "Is salt an element?"—*No*. "Is honey an element?"—*No*. "Is iron an element?"—*Yes*. But it is clear that this wouldn't get us very far. We would soon run into big puzzles. "Are there humors?"—*No*. "Do choleric people live shorter lives?"—*Yes*. Even having all of our envisaged scientific questions answered will not carry us all that far toward securing an adequate science. For in forming our questions we must operate within the concept framework of the day. (How could it be otherwise?) And the prospect of inadequacy roots in just this circumstance. Getting our questions answered might not mean all that much. The important innovations of science are always conceptual innovations, and the conceptual innovations of the future lie by definition beyond our present horizons.

The erotetic completeness of a state of the cognitive art may merely reflect its blindness: even if we could realize the erotetic completeness of our knowledge, we could not be confident that we have actually brought the project of inquiry to a satisfactory conclusion. Erotetic completeness may well indicate poverty rather than wealth. Even as its capacity to resolve our questions counts as a merit of a body of knowledge, so does its capacity to raise new questions of significance and depth. The opening up of new lines of investigation and research qualifies among the prime strengths of a new state of the art in scientific inquiry, and ranks, as such, among its virtues. Completeness is the hallmark of a pseudoscience, which is generally so contrived that the questions that are allowed to be raised are exactly the questions the projected machinery is in a position to resolve.

But, in any case, erotetic completeness is simply unachievable for natural science, which is characterized by Kant's Principle of Question Propagation. Science emerges as a project of self-transcendence. It embodies an inner drive that always presses beyond the capacity limits of the historical present.[16]

2. Questions and Scientific Progress

Synopsis
(1) Consideration of various theories of scientific progress in the light of the cognitive dialectic of questions and answers suggests that no purely theoretical criterion of progressiveness is ever adequate. (2) Moreover, history teaches the dangers of drawing conclusions from the perceived completeness of science. (3) We cannot, without circularity, appraise the progress of theory by purely theoretical standards but must step outside the realm of pure theory into that of practice.

1. Question Dialectics and Scientific Progress

Various accounts have been proposed to characterize scientific progress in terms of historical tendencies regarding question-and-answer relationships. Perhaps the most rudimentary theory of this sort is the traditional *cumulationist* view that later, more advanced stages of science are characterized as such by virtue of their *answering more questions*—questions over and above those answered at earlier stages of the game:

$$t < t' \supset [\mathbf{Q}^*(S_t) \subset \mathbf{Q}^*(S_{t'})]$$

(Here $\mathbf{Q}^*(S_t)$ is the set of all S_t-answered questions, comprising all of those S_t-posable questions for which S_t also provides an answer.) This expansionist view of progress has it that later, superior science answers all of the formerly answered questions (albeit perhaps differently) and, furthermore, answers some previously unanswered questions. Progress, according to this theory, is a matter of rolling-snowball–analogous knowledge accumulation: as science progresses, the set of answered questions is an ever-growing whole.

Along this line, Karl R. Popper has suggested[1] that if the "content" of a scientific theory T is construed as the set of all questions to which it can provide answers, then a scientific theory might be compared unfavorably with that of its superior successors—notwithstanding their substantive difference—because of proper inclusion with respect to this erotetic mode of "content." Popper asserted that, on this view, even though T_1 is assertorically incompatible with T_2, we might still be in a position to compare the question sets $\mathbf{Q}^*(T_1)$ and $\mathbf{Q}^*(T_2)$ and, in particular, might have the inclusion relation $\mathbf{Q}^*(T_1) \subset \mathbf{Q}^*(T_2)$.

But this is nonsense. For let p be a proposition that brings this assumed incompatibility to view, so that T_1 asserts p and T_2 asserts not-p. Then "Why is p the case?" is a question that the theory T_1 not only allows to arise but also presumably furnishes with an answer. So this question belongs to $\mathbf{Q}^*(T_1)$. But it cannot belong to $\mathbf{Q}^*(T_2)$, because T_2 *(ex hypothesi)* violates the question's evident presupposition that p is the case.[2]

If one body of assertions includes the thesis p among its entailments, while another fails to include p, then we can always ask (as above) the rationale-demanding question, "Why is it that p is the case?" with respect to the former, where the presupposition that p is the case is met, but now with respect to the latter, where this presupposition fails. Discordant bodies of (putative) knowledge engender distinct, mutually divergent bodies of questions because they provide the material for distinct background presuppositions. Modern medicine no longer asks about the operations of Galenic humors; present-day physics no longer asks about the structure of the luminiferous ether; future physics may well no longer ask about the characteristics of quarks. When science abandons certain theoretical entities, it also forgoes (and gladly forgoes) the opportunity of asking questions about them.

In the actual course of scientific progress, we see not only gains in question resolution but losses as well. Aristotle's theory of natural place provided an explanation for the "gravitational attraction" of the earth in a way that Newton's theory did not. Descartes's vortex theory could answer the question of why all the planets revolve about the sun in the same direction, a question to which Newton's celestial mechanics had no answer. The earlier chemistry of affinities had an explanation for why certain chemical interactions take place and others not, a phenomenon for which Dalton's quantitative theory had no explanation. The cumulationist theory of progress through an ongoing enlargement in question resolution is patently untenable.

A second theory of progress takes the rather different approach of associating scientific progress with an expansion in our question horizon. It holds that later, superior science will always enable us to pose additional questions:

$t < t' \supset [\mathbf{Q}(S_t) \subset \mathbf{Q}(S_{t'})]$

Scientific progress is now seen as a process of enlarging the question agenda by uncovering new questions. More questions rather than more answers are seen as the key: progress is a matter of question cumulativity, with more advanced science making it possible to pose issues that could not be envisaged earlier on.

This second approach to progressiveness is also untenable. For just as later and better science sometimes involves abandoning old answers, so it also often involves rejecting old questions. Paul Feyerabend has argued this point cogently. New theories, he holds, generally do not subsume the issues of the old but move off in altogether different directions. At first, the old theory may even be more comprehensive—having had more time for its development. Only gradually does this alter. But by then "the slowly emerging conceptive apparatus of the [new] theory soon starts defining its own problems, and earlier problems, facts, and observations are either forgotten or pushed aside as irrelevant."[3]

A third theory sees scientific progress in terms of an increase in the volume of resolved questions. Thus, Larry Laudan has argued against Popper that scientific progress is not to be understood as arising because the later, superior theories that replace our earlier ones answer all the questions answered by their rival (or earlier) counterparts, plus some additional questions, but rather simply because the replacing theories answer *more* questions (although not necessarily all of the same questions answered previously).[4] On such a doctrine, progress turns on *a numerical increase in the sheer quantity of answered questions:*

$$t < t' \supset [\#\mathbf{Q}^*(S_t) < \#\mathbf{Q}^*(S_{t'})]$$

This position also encounters grave difficulties. For how are we going to do our bookkeeping? How can we individuate questions for the counting process? Just how many questions does "What causes cancer?" amount to? And how can we avoid the ambiguity inherent in the fact that once an answer is given, we can always raise further questions about its inner details and its outer relationships?

Moreover, this position is unpromising as long as we leave the *adequacy* of the answers out of account. In its earliest, animistic stage, for example, science had answers for everything. Why does the wind blow? The spirit of the winds arranges it. Why do the tides ebb and flow? The spirit of the seas sees to it. And so on. Or again, take astrology. Why did X win the lottery or Y get killed in an accident? The conjunction of the stars provides all the answers. Some of the biggest advances in science come about when we reopen questions—when our answers get unstuck en masse with the discovery that we have been on the wrong track, that we do not actually understand something we thought we understood perfectly well.

A fourth theory sees scientific progress in terms of a decrease in the sheer number of unanswered questions:

$$t < t' \supset \{[\#Q(S_t) - \#Q^*(S_t)] < [\#Q(S_{t'}) - \#Q^*(S_{t'})]\}$$

This approach sees progress as a matter of agenda diminution. Progressiveness turns on a numerical decrease in the register of unanswered questions. But the same line of objection put forth against its predecessor will also tell against this present conception of progress. To be sure, the size of this gap between Q and Q* is something significant—a measure of the *visible inadequacy* of a given state of the cognitive art. The striving to close this gap is a prime mover of scientific inquiry, but it is emphatically not an index of progress.

An increase in the volume of *unanswered* questions is compatible with a more than compensating increase in the volume of *answered* questions. A fifth cognate theory accordingly sees progressiveness in terms of a decrease in the relative proportion of answered questions and uses the ratio of answered to unanswered questions as a touchstone:

$$t < t' \supset \left[\frac{\#Q^*(S_t)}{\#Q(S_t)} < \frac{\#Q^*(S_{t'})}{\#Q(S_{t'})} \right]$$

Just why should one think this relationship to be essential to progress? It is perfectly possible, in theory, that scientific progress might be divergent, that particular increases in question-resolving capability might be more than offset by expanding problem horizons, in roughly the manner of figure 2.1. (10 percent of 10^7 questions is still a substantially bigger number than 20 percent of 10^6 questions.) In such circumstances, we could make striking "progress" by way of substantial increases in question-resolving capacity, but we would nevertheless have a smaller proportion of answered questions because of the larger volume of new questions.

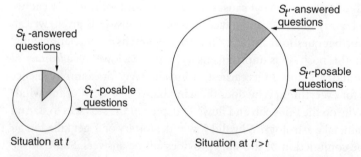

Figure 2.1

Some theoreticians have favored yet another, sixth theory—one that sees scientific progress as essentially ignorance-enlarging. That is, they regard scientific progress as a matter of decreasing the relative proportion of answered questions. Thus, W. Stanley Jevons wrote:

> In whatever direction we extend our investigations and successfully harmonise a few facts, the result is only to raise up a host of other unexplained facts. Can any scientific man venture to state that there is less opening now for new discoveries than there was three centuries ago? Is it not rather true that we have but to open a scientific book and read a page or two, and we shall come to some recorded phenomenon of which no explanation can yet be given? In every such fact there is a possible opening for new discoveries.[5]

This theory sees scientific progress as a cognitively divergent process, subject to the condition that the more we know, the more we are brought to the realization of our relative ignorance. But this position also has a serious flaw. It totally fails to do justice to those more routine stages of the history of science—by no means infrequent—when progress does go along in the manner of the classical pattern of an increase in both the volume and the proportion of resolved questions.

In sum, none of these approaches to progress through question-answer relationships seems promising.

2. The Lessons of History

In relating $Q(S_t)$ and $Q^*(S_t)$ in the manner of the progress theories of the preceding section, we operate merely in the realm of appearances—how far the (putative) science of the day can go in resolving the visible problems of the day. This whole approach is too fortuitous and context-dependent to bear usefully on anything so fundamental. Apparent adequacy relative to the existing body of knowledge (which, after all, is the best we can do in this direction) is a very myopic guide. Historical considerations are instructive in this regard.

Think here of the situation that prevailed in the last quarter of the nineteenth century. Lord Kelvin thought that the basic structure of physics was pretty well completed and that no really major questions remained to be answered. In 1885, Berthelot maintained that "the world is now without mysteries."[6] In his popular text on electricity (1887), T. C. Mendenhall, formally a physics professor, at the time a college president, and soon to be both president of the American Association for the Advancement of Science and superintendent of the United States Coast and Geodetic Survey, maintained that:

> More than ever before in the history of science and invention, it is safe now to say what is possible and what is impossible. No one would claim for a moment that during the next five hundred years the accumulated stock of knowledge of geography will increase as it has during the last five hundred.... In the same way it may safely be affirmed that in electricity the past hundred years is not likely to

be duplicated in the next, at least as to great, original, and far-reaching discoveries, or novel and almost revolutionary applications.[7]

An even more remarkable instance of the same phenomenon is given by the strikingly parallel report of Max Planck:

> As I was beginning to study physics [in 1875] and sought advice regarding the conditions and prospects of my studies from my eminent teacher Philip von Jolly, he depicted physics as a highly developed and virtually full-grown science, which—since the discovery of the principle of the conservation of energy had in a certain sense put the keystone in place—would soon assume its finally stable form. Perhaps in this or that corner there would still be some minor detail to check out and coordinate, but the system as a whole stood relatively secure, and theoretical physics was markedly approaching that degree of completeness which geometry, for example, had already achieved for hundreds of years. Fifty years ago [as of 1824] this was the view of a physicist who stood at the pinnacle of the times.[8]

Such sentiments represent a then widespread tendency. The very fact that the most "advanced" science of the day—physics—seemed to be nearing the end of the line, together with a contemplation of the enormous strides being made all across the scientific frontier—in biology, medicine, chemistry, and so on—opened up the seemingly plausible prospect that science stood pretty much at the last frontiers. The course of progress in scientific knowledge—so dramatically explosive since its first great flourishing in the seventeenth century—now seemed to be nearing its final and completed stage. The ethos of science during the closing quarter of the last century was one of success and self-congratulation. The dominant sentiment in metascientific theory was one of elation, of pride in the face of immense strides—feelings of power approaching hubris that the intellectual conquest of nature was virtually complete.[9]

This self-congratulatory spirit of admiration of a completed job well done was a transient phenomenon, however. As Rutherford's teacher, J. J. Thomson, observed in his presidential address to the British Association in 1909:

> The new discoveries made in physics in the last few years, and the ideas and potentialities suggested by them, have had an effect upon the workers in that subject akin to that produced in literature by the Renaissance. Enthusiasm has been quickened, and there is a hopeful, youthful, perhaps exuberant, spirit abroad which leads men to make with confidence experiments which would have been thought fantastic twenty years ago. It has quite dispelled the pessimistic feeling, not uncommon at that time, that all the interesting things had been discovered, and all that was left was to alter a decimal or two in some physical constant. There never was any justification for this feeling, there never were any

signs of an approach to finality in science. The sum of knowledge is at present, at any rate, a diverging not a converging series.[10]

Beginning in the first decade of the twentieth century, the outlook on the prospect of the physical sciences changed dramatically, and soon a wholly new outlook prevailed. The reason for this change of view is all too plain when one considers the list of the Nobel prize winners in physics during the era from Röntgen to Yukawa.[11] The revolution wrought by these men in our understanding of nature was so massive that their names became household words throughout the scientifically literate world. Nor was progress confined to the sphere of theory. By the end of World War II, physics had reached awesome heights of technological impact in the atom bomb and the prospects of peaceful uses of nuclear power that lay just around the corner. Given this, it was inevitable that a drastic alteration in informed views of the prospects of science should have come about. By the 1920s, this phenomenon was well underway, and it reached its heyday in the years immediately following World War II.

Recently, however, the pendulum has swung back. Thus, Richard Feynman, the American 1965 Nobel physics laureate who is by any reckoning among the best physicists of the day, has written as follows:

> What of the future of this adventure? What will happen ultimately? We are going along guessing the laws; how many laws are we going to have to guess? I do not know. Some of my colleagues say that this fundamental aspect of our science will go in; but I think there will certainly not be perpetual novelty, say for a thousand years. This thing cannot keep on going so that we are always going to discover more and more new laws.... We are very lucky to live in an age in which we are still making discoveries. It is like the discovery of America—you only discover it once. The age in which we live is the age in which we are discovering the fundamental laws of nature, and that day will never come again. It is very exciting, it is marvelous, but this excitement will have to go. Of course in the future there will be other interests,... but there will not still be the same things that we are doing now.... There will be a degeneration of ideas, just like the degeneration that great explorers feel is occurring when tourists begin moving in on a territory. In this [present] age people are [perhaps for the last time?!] experiencing a delight, the tremendous delight that you get when you guess how nature will work in a situation never seen before.[12]

Sentiments of this sort abound in the recent literature and can be cited from writers of virtually every level of scientific sophistication.

One particularly interesting and well-developed statement is that of the biologist Gunther S. Stent in *The Coming of the Golden Age: A View of the End of Progress*. The following quotation will convey the flavor of Stent's discussion:

> I want to consider what I believe to be intrinsic limits to the sciences, limits to the accumulation of meaningful statements about the events of the outer world. I think everyone will readily agree that there are *some* scientific disciplines which, by reason of the phenomena to which they purport to address themselves, are *bounded*. Geography, for instance, is bounded because its goal of describing the features of the Earth is clearly limited. . . . And, as I hope to have shown in the preceding chapters, genetics is not only bounded, but its goal of understanding the mechanism of transmission of hereditary information *has*, in fact, been all but reached. . . . [To be sure] the domain of investigation of a bounded scientific discipline may well present a vast and practically inexhaustible number of events for study. But the discipline is bounded all the same because its goal is in view. . . . There is at least one scientific discipline, however, which appears to be *open-ended*, namely physics, or the science of matter. . . . But even though physics is, in principle, open-ended, it too can be expect to encounter limitations in practice,. . . [for] there are purely physical limits to physics because of man's own boundaries of time and energy. These limits render forever impossible research projects that involve observing events in regions of the universe more than ten or fifteen billion light-years distant, traveling far beyond the domain of our solar system, or generating particles with kinetic energies approaching those of highly energetic cosmic rays.[13]

As one reads these discussions, one gets the feeling of having seen it all before—as indeed one has. On the issue of the perceived completeness of science, the pendulum of fashion swings back and forth with the ever-changing appearances of the landscape of the moment.

The lesson of these historical considerations is simply that the perceived adequacy of science reflected in the relationship of $Q(S_n)$ to $Q^*(S_n)$ is a rollercoaster that affords little useful insight into the fundamentals of scientific completeness. The testimony of history illustrates the findings of our theoretical analysis in this regard.

3. The Pragmatic Dimension of Progress

It is time to step back from the proliferation of doctrines relating progress to questions and view the matter in a wider perspective. The salient point is that even if the historical course of scientific inquiry had in fact conformed, overall, to one or another of the patterns of question-answer dialectic envisaged by these various theories, this circumstance would simply be fortuitous. It would not reflect any deep principle inherent in the very nature of the enterprise. For progress as such does not hinge on any such comparisons of the relative sizes of question sets and answer sets.

A common failing of all the approaches discussed above is that they deal (in the first instance, at any rate) simply with questions as such, without worrying about their significance. To render such theories at all meaningful, this factor would have to be reckoned with. The theory would have to be construed as applying not to questions per se but to *important* questions—questions at or above some suitable level of significance.

At this stage, this entire program of determining progressiveness in terms of question-answer relations runs into fatal difficulties. To see this, let us consider the issue of the significance of questions somewhat more closely.

Clearly, one question can *include* another, as "What causes lightning?" includes "What causes ball lightning?" In the course of answering the one, we are called on to provide an answer for the other. Such relations of inclusion and dominance provide a basis for comparing the "scope" of questions (in one sense of this term) in certain cases—although certainly not in general. They do not enable us to compare the scope of "What causes lightning?" and "What causes tides?" And even if we could (per impossible) measure and compare the size of questions in this sort of (content-volume oriented) sense, this would afford no secure guide to their relative importance.

The importance of a factual question Q, where $Q \in \mathbf{Q}(S)$, turns in the final analysis on how substantial a revision in our body of scientific beliefs S is wrought by our grappling with it, that is, the extent to which answering it causes geological tremors across the cognitive landscape. Two very different sorts of things can be at issue here: either a mere *expansion* of S by additions, or, more seriously, a *revision* of it that involves replacing some of its members and readjusting the remainder so as to restore overall consistency. This second sort of change in a body of knowledge, its revision rather than mere augmentation, is in general the more significant matter, and a question whose resolution forces revisions is likely to be of greater significance than one that merely fills in some part of the terra incognita of knowledge. Obviously, the magnitude of the transformation from an earlier S to a later successor S' can only be assessed once we have arrived at S'.

Accordingly, the importance or interest of a question is something that can only be discovered with hindsight from the vantage point of the new body of knowledge S', to which the attempt to grapple with S had led us. In science, apparently insignificant problems (the blue color of the sky, or the anomalous excess of background radiation) can acquire great importance once we have a state of the art that makes them instances of important new effects—effects that provide guideposts toward major theoretical innovations.

The crucial fact is that progressiveness, insignificance, importance, interest, and the like are all state of the art-relative conceptions. To apply these ideas, we must already have a particular scientific corpus in hand to serve as a vantage point for their assessment. No commitment-neutral basis is available for deciding whether S_1 is progressive vis-à-vis S_2 or the reverse. If the test of a theory is to be its problem-solving capacity—its capacity to provide viable answers to *interesting* questions[14]—then this is something in which the theory itself (as an integral component of the scientific state of the art) is going to play a pivotal role. This is not an ultimately happy state of affairs because of the circularity it engenders. We had best look in a very different direction for the standard of progressiveness.

The most promising prospect here calls for approaching the issue of scientific progress in terms of pragmatic rather than strictly cognitive standards. Progressively superior science does not manifest itself as such through the sophistication of its theories (for, after all, even absurd theories can be made very complex), but through the superiority of its applications as judged by the old Bacon-Hobbes standard of *scientia propter potentiam*—increased power of prediction and control. In the end, praxis is the arbiter of theory.[15] To understand scientific progress and its limits, we must look not toward the dialectic of questions and answers but toward the scope and limits of human power in our transactions with nature.

The Instability of Science

Synopsis
(1) Science is a comparatively fragile structure: the best it can do is provide ever-changing estimates of how things work. Our scientific knowledge is always defeasible and transitory. The scientific theory of one day is destined to be rejected by that of the next. (2) At the level of scientific theories, our commitments are tentative and fallible—in sum, provisional. (3) We must adopt the stance of a cognitive Copernicanism that abandons the idea that we occupy a privileged historical position in the course of scientific development. Our science is nothing permanent—it is (at most) the best that can be done at this point, in this particular state of the art. As with any human artifact, a scientific theory is fated from the outset to be swept away by the ravages of time. (4) Its instability and changeability notwithstanding, natural science does indeed progress—not, to be sure, by reaching (or even approaching) "the ultimate truth" but by providing us with increasingly powerful instrumentalities for successful prediction and control.

1. The Comparative Fragility of Science: Scientific Claims as Mere Estimates

Scientific "knowledge" at the level of deep theory is always purported knowledge: knowledge as we see it today. In our heart of hearts, we realize that we may see it differently tomorrow—or the day after. We must stand ready to acknowledge the fragility of our scientific theorizing. All we are ever able to do in natural science is to select the optimal answer to the questions we manage to formulate within the realm of alternatives specifiable by means of the conceptual machinery of the day. And we have no reason to doubt—nay, we have every reason to believe—that the day will come when this conceptual basis will be abandoned, in the light of yet unrealizable developments, as altogether inadequate.

Give that S_n is *(ex hypothesi)* our own current body of (putative) scientific knowledge—the science of the day, S_t with $t = n$ for "now"—we ourselves unquestionably stand committed to the inference schema: if $p \in S$, then p. At the level of specific claims, we have little alternative but to look upon *our* knowledge as *real* knowledge—a thesis p would not be part of "*our* truth" if we did not *take* it to form part of "*the* truth." The answers we give to our questions are literally the best we can provide. The "knowledge" at issue is knowledge according to our own lights—the only ones we've got. Despite our resort to this adequacy principle at the level of particular claims, we nevertheless should not have the hubris to think that our science has got it right.

The relationship of our (putative) scientific knowledge to the so-called (real) truth has to be conceived of in terms of estimation. At the frontiers of generality and precision, our truth in matters of scientific theorizing is not—and may well never actually be—the real truth. Science does not secure the truth (deliver it into our hands in its definitive finality). We have no alternative to acknowledging that our science, as it stands here and now, does not present the real truth to provide us with a tentative and provisional *estimate* of it. However confidently it may affirm its conclusions, the realization must be maintained that the declarations of natural science are provisional and tentative—subject to revision and even to outright rejection. The most we can ever do is to take our science (S_n) as the imperfect best we can do here and now to conjecture "the real truth."

To be sure, there is no question that we can improve our science. The history of science cries out for the Whig interpretation. Every applicable standard, from systemic sophistication to practical applicability, yields reason to think that later science is better science. Of course, this does not mean that later science is any the truer—for true it cannot be if it, too, is destined for eventual rejection. The standards of scientific acceptability do not and cannot assure actual or indeed even probable or approximate truth. As Larry Laudan has argued with substantial evidence and eloquence: "No one has been able even to say what it would mean to be 'closer to the truth,' let alone to offer criteria for determining how we could assess such proximity."[1] All claims that emerge from scientific theorizing are vulnerable—subject to improvement and replacement. We can claim that later, "superior" science affords a *better warranted estimate* of the truth, but we cannot claim that it manages somehow to capture more of the truth or to approximate to it more closely.[2]

Interestingly enough, in point of vulnerability science compares unfavorably with the commonplace knowledge of matters of prescientific fact—our knowledge regarding sticks and stones and sealing wax and the other paraphernalia of everyday life. And there is good reason for this.

Increased security can always be purchased for our estimates at the price of decreased accuracy. We estimate the height of a tree at around 25 feet. We are *quite sure* that the tree is 25 ± 5 feet high. We are *virtually certain* that its height is 25 ± 10 feet. But we are *completely and absolutely sure* that its height is between 1 inch and 100 yards. Of this we are "completely sure" in the sense that we are "absolutely certain," "certain beyond the shadow of a doubt," "as certain as we can be of anything in the world," "so sure that we would be willing to stake our life on it," and the like. For any sort of estimate whatsoever, there is always a characteristic tradeoff between the evidential security or reliability of the estimate, on the one hand (as determinable on the basis of its probability or degree of acceptability), and, on the other hand, its contentual *definiteness* (exactness, detail, precision, etc.). A situation of the sort depicted by the concave curve of figure 3.1 obtains. (Compare also chapter 7, below.)

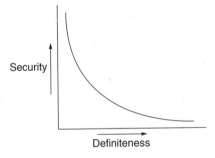

Figure 3.1 The Degradation of Security with Increasing Definiteness

Now, the crucial point is that natural science eschews the security of indefiniteness. In science we operate at the righthand side of the diagram: we always strive for the maximal achievable universality, precision, exactness, and so on. The law claims of science are strict: precise, wholly explicit, exceptionless, and unshaded. They involve no hedging, no fuzziness, no incompleteness, and no exceptions. In stating that "the melting point of lead is 327.545 °C at standard pressure," the physicist asserts that all pieces of (pure) lead will unfailingly melt at exactly this temperature; he certainly does not mean to assert that most pieces of (pure) lead will probably melt at somewhere around this temperature. By contrast, when we assert in ordinary life that "peaches are delicious," we mean something like "most people will find the eating of suitably grown and duly matured peaches a relatively pleasurable experience." Such statements have all sorts of built-in safeguards like "more or less," "in ordinary circumstances," "by and large," "normally," "if other things are equal," and the like. They are not laws but rules of thumb, a matter of practical lore rather than scientific rigor. In natural science, however, we deliberately accept risk by aiming at maximal definiteness—and thus at maximal informativeness and testability. The theories of natural science take no notice of what happens ordinarily or normally; they seek to transact their explanatory business in terms of high generality and strict universality—in terms of what happens always and everywhere and in all circumstances. In consequence, we must recognize the vulnerability of our scientific statements. The fact that the theoretical claims of science are mere estimates that are always cognitively at risk and enjoy only a modest life span has its roots in science's inherent commitment to the pursuit of maximal definiteness. Its cultivation of informativeness (of definiteness of information) everywhere forces science to risk error.

A tenable fallibilism must take this inverse relationship between definiteness and security into careful account. Consider the "atomic theory," for example. It has an ancient and distinguished history in the annals of science, stretching from the speculations of Democritus in antiquity, through the work of Dalton, Rutherford, and Bohr, to the baroque complexities of the present day. "It is surely unlikely that science will ever give up on atoms!" you say. Quite true! But what we are dealing with here is clearly not a single scientific theory at all, but a vast family of scientific theories, a great bundle loosely held together by threads of historical influence and family resemblances. There is not much that Rutherford's atoms, and Bohr's, and our contemporary quantum theorists' have in common. As such, the atomic theory is no more than a rough generic schema based on the more or less metaphorical intuition that "matter is granular in the small, composed of tiny structures separated in space." This is surely incomplete and indeterminate—a large box into which a vast number of particular theories can be fitted. It claims that there are atoms but leaves open an almost endless range of possibilities as to what they are like. This sort of contention may well be safe enough; at this level of schematic indiscriminateness and open-endedness, scientific claims can of course achieve security. But they do so only at the expense of definiteness—of that generality and precision that reflect what science is all about.

The quest for enhanced definiteness is unquestionably a prime mover of scientific inquiry. The ever-continuing pursuit of increasing accuracy, greater generality, widened comprehensiveness, and improved systematicity for its assertions is the motive force behind scientific research. This innovative process—impelled by the quest for enhanced definiteness—drives the conceptual scheme of science to regions ever more distant from the familiar conceptual scheme of our everyday life. For the ground rules of ordinary life discourse are altogether different. Ordinary life communication is a practically oriented endeavor carried on in a social context: it stresses such maxims as "Aim for security, even at the price of definiteness"; "Protect your credibility"; "Avoid misleading people, or—even worse—lying by asserting outright falsehoods"; "Do not take a risk and 'cry wolf.'"

The aims of ordinary life discourse are primarily practical, largely geared to social interaction and the coordination of human effort. In this context, it is crucial that we aim at credibility and acceptance—that we establish and maintain a good reputation for reliability and trustworthiness. In the framework of common life discourse, we thus take our stance at a point far removed from that of science. Very different probative orientations prevail in the two areas. In everyday contexts, our approach is one of *situational satisficing*: we stop at the first level of sophistication and complexity that suffices for our present needs. In science, however, our objectives are primarily theoretical and governed by the aims of disinterested inquiry. Here our approach is one of systemic maximizing: we press on toward the ideals of systemic completeness and comprehensiveness. In science we put ourselves at greater risk because we ask much more of the project.

A view along the following lines is very tempting: "Science is the best, most thoroughly tested knowledge we have; the 'knowledge' of everyday life pales by comparison. The theses of science are really secure; those of everyday life, casual and fragile." The very reverse is actually the case: our scientific theories are vulnerable and have a short life span; it is our claims at the looser level of ordinary life that are relatively secure and stable.

Of course, the claims of everyday life have sacrificed definiteness in order to gain this security and reliability. One recent writer has quite correctly written that:

> we sometimes forget ... how completely our own knowledge has absorbed what the cavemen knew, and what Ptolemy, Copernicus, Galileo and Newton knew. True, some of the things our predecessors thought they knew turned out not to be knowledge at all: and these counterfeit facts have been rejected along with the theories that "explained" them. But consider some of the facts and low-level regularities our ancestors knew, which we know, and which any astronomical theory is obligated to explain: the sky looks about the same every night; it is darker at night than during the day; there is a moon; the moon changes its appearance regularity; there is a cycle of seasons.[3]

To be sure, the sort of knowledge at issue here does not go very far in affording us a detailed understanding of the ways of the world. It lacks detailed scientific substance; but just exactly that is the basis of its comparative security.

We must recognize, without lapsing into skepticism, that the cognitive stance characteristic of science requires the acceptance of fallibility and corrigibility, and so requires a certain tentativeness engendering the presumption of error. Because the aims of the enterprises are characteristically different, our inquiries in everyday life and in science have a wholly different aspect; the former achieves stability and security at the price of sacrificing definiteness; a price that the latter scorns to pay.

2. Fallibilism and the Distinction Between Our (Putative) Truth and the Real Truth

We would like to think of our science as money in the bank—as something safe, solid, and reliable. Unfortunately, however, history militates against this comfortable view of our scientific theorizing. In science, new knowledge does not just supplement but generally upsets our knowledge-in-hand. We must come to terms with the fact that—at any rate, at the scientific level of generality and precision—*each* of our accepted beliefs *may* turn out to be false, and many of our accepted beliefs *will* turn out to be false.

Let us explore the ramifications of a cognitive fallibilism that holds that science—our science, the body of our scientific beliefs as a whole—consists largely, and even predominantly, of false beliefs, embracing various theses that we will ultimately come to see (with the wisdom of hindsight) as quite untenable. It is useful to distinguish between potential error at the particularistic and collectivistic levels—between thesis defeasibility and systemic fallibilism. The former position makes the claim that *each* of our scientific beliefs *may* be false, while the latter claims that *many* or *most* of our scientific beliefs *are* false. The two theses are independent. Of the pair, "Each might be false" and "Most are false," neither entails the other. (Each of the entrants in the contest might be its winner, but it could not turn out that most of them indeed are. Conversely, most of the integers of the sequence 1, 2, 3, ... come after 3, but it cannot be that all might do so.)

Although the theses are independent, both are true. Thesis defeasibility follows from the fact that it is perfectly possible—and not only theoretically but realistically possible—that one could play the game of scientific inquiry correctly by all the accepted rules and still come to a result that fails to be true. For the truth of our objective claims hinges on how matters really stand in the world and not on the (inevitably incomplete) grounding or evidence of justification that we ourselves have in hand. We cannot cross the epistemic gap between the apparent and the real by any logically secure means; and so we must always be careful to distinguish between "our (putative or ostensible) truth," what we think to be true, on the one hand, and on the other, "the real truth," what actually is true—the authentic and unqualified truth of the matter. In scientific theorizing, we are never in a position to take this epistemic gap as closed.

Systemic fallibilism is the more serious matter, however, for it means that one must presume that our whole scientific picture of the world is seriously flawed and will ultimately come to be recognized as such—that most of our present-day beliefs at the scientific frontier will eventually have to be abandoned. We learn by empirical inquiry about empirical inquiry, and one of the key things we learn is that at no actual stage does science yield a final and unchanging result. All the experience we can muster indicates that there is no justification for regarding our science as more than an inherently imperfect stage within an ongoing development. And we have no responsible alternative to supposing the imperfection of what we take ourselves to know. We occupy the predicament of the Preface Paradox, exemplified by the author who apologizes in his preface for those errors that have doubtless made their way into his work and yet blithely remains committed to all those assertions he makes in the body of the work itself. We know or must presume that (at the synoptic level) there are errors, although we certainly cannot say where and how they arise. There is no realistic alternative to the supposition that science is wrong in various ways and that much of our supposed "knowledge" of the world is no more than a tissue of plausible error. We are thus ill-advised to view the science of our day—or of any day—as affording the final truth of the matter.

It goes without saying that we can never be in a position to claim justifiedly that our current corpus of scientific knowledge has managed to capture "the whole truth." We are also not even in a position justifiedly to claim that we've got "nothing but the truth." Our scientific picture of nature must always be held provisionally and tentatively, however deeply we may be attached to some of its details.

All attempts to equate the real truth with apparent truth of some sort are condemned to failure: having the same generic structure, they have the same basic defect. They all view "the real truth" as "the putative truth when arrived at under conditions C" (be it in favorable circumstances or under careful procedures). And in each case the equation is destroyed by the ineliminable possibility of a separation between "the real truth" and the putative truth that we ourselves actually reach (by whatever route we have arrived there). None of our scientific beliefs are so cataleptic or so clear and distinct that they cannot run awry. Nobody's putative scientific truth can be authenticated as genuine—be it that of the experts or of those who prudently implement established epistemic methods. With claims to knowledge at the level of scientific generality and precision, there is no way in which we can effect an inferentially secure transition to the real truth from anybody's ostensible truth (our own included).

The transition from appearance to reality can perhaps be salvaged by taking God or the community of perfect scientists or some comparable idealization to be the "privileged individual" whose putative truth is at issue. The doctrine can thus be saved by resorting to the Myth of the God's Eye View: with S^*

as the cognitive corpus of "perfected science," we can indeed establish an equivalence between S^* membership and "the real truth." But this desperate remedy is of little practical avail. For only God knows what "perfected science" is like; we ourselves certainly do not. And the schoolmen were right: in these matters we have no right to call on God to bail us out. (*Non in scientia recurrere est ad deum.*) While the idea at issue may provide a useful contrast or conception, it is something that we cannot put to concrete implementation.

Of course, we stand committed to our truths, imperfect though they may be; but we have nowhere else to go. We have no alternative to proceeding on the working hypothesis that in scientific matters *our* truth is *the* truth. But we must also recognize that that is simply not so—that the working hypothesis in question is no more than just that, a convenient fiction.

We recognize that the workaday identification of our truth with *the* truth is little more than a makeshift—albeit one that stands above criticism because no alternative course is open to us. The recognition that this may well not be good enough is simply an ineliminable part of the human condition. No deity has made a covenant with us to assure that what we take to be adequately established theories must indeed be true in these matters of scientific inquiry.

The proper lesson here is not the skeptic's conclusion that our science counts for naught, but simply, and less devastatingly, the realization that our scientific knowledge—or purported knowledge—of how things work in the world is flawed; that it is replete with errors that we ourselves are impotent to distinguish from the rest. With respect to the claims of science, we realize that our herd is full of goats that nevertheless appear altogether sheeplike to us.

3. Cognitive Copernicanism

The fallibility and corrigibility of our science means it cannot be viewed as providing definitive (let alone absolutely true) answers to its questions. We have no alternative but to see *our* science as both incomplete and incorrect in some (otherwise unidentifiable) respects. In this sense—that of its inability to deliver into our hands something that can be certified as the truth, the whole truth, and nothing but the truth—science is certainly subject to a severe limitation. What we proudly vaunt as scientific knowledge is a tissue of hypotheses—of tentatively adopted contentions, many or most of which we will ultimately come to regard as quite untenable and in need of serious revision or perhaps even rejection.

If there is one thing we can learn from the history of science, it is that the scientific theorizing of one day is looked upon by that of the next as deficient. The clearest induction from the history of science is that science is always mistaken—that at every stage of its development, its practitioners, looking backward with the wisdom of hindsight, will view the work of their predecessors as seriously deficient and their theories as fundamentally mistaken. If we adopt (as in candor we must) the modest view that we ourselves and our contemporaries do not occupy a privileged position in this respect, then we have no reasonable alternative but to suppose that much or all of what we ourselves vaunt as "scientific knowledge" is itself presumably wrong.[4]

We must stand ready to acknowledge the fragility of our scientific knowledge and must temper our claims to such knowledge with a cognitive Copernicanism. The original Copernican revolution made the point that there is nothing ontologically privileged about our own position in space. The doctrine now at issue effectively holds that there is nothing cognitively privileged about our own position in time. It urges that *there is nothing epistemically privileged about the present—any present*, our own prominently included. Such a perspective indicates not only the incompleteness of our knowledge but its presumptive incorrectness as well. The current state of knowledge is simply one state among others, all of which stand on an imperfect footing.

No human generation is cognitively advantaged. We must acknowledge that the transience that has characterized all scientific theories of the historical past will possibly, even probably, characterize those of the present as well. Given the historical realities, the idea that science does—or sooner or later must—arrive at a changeless vision of "the truth of the matter" is not plausible. We have no alternative but to reject the egocentric claim that we ourselves occupy a pivotal position in the epistemic dispensation, and we must recognize that there is nothing inherently sacrosanct about our own present cognitive posture vis-à-vis that of other, later historical junctures. A kind of intellectual humility is in order—a diffidence that abstains from the hubris of pretentions to cognitive finality or centrality. Such a position calls for the humbling view that just as we think our predecessors of a hundred years ago had a fundamentally inadequate grasp on the furniture of the world, so our successors of a hundred years hence will take the same view of our purported knowledge of things. Realism requires us to recognize that, as concerns our scientific understanding of the world, our most secure knowledge is very likely no more than presently accepted error.

To be sure, this recognition of the fallibilism of our cognitive endeavors must emphatically *not* be construed as an open invitation to a skeptical abandonment of the cognitive enterprise. Instead, it is an incentive to do the very best we can. In human inquiry, the cognitive ideal is correlative with the striving for optimal systematization. And this is an ideal that, like other ideals, is worthy of pursuit, despite the fact that we must recognize that its full attainment lies beyond our grasp.

The crucial consideration thus remains that science is not and presumably never can be in a position to offer us anything that is definitive, incorrigible, and final. Our science is and must be developed within the confines of man's cognitive fallibility. He who looks to science for answers that are ultimate and absolute is destined to look in vain. We must recognize that "our science" is not something permanent, secured for the ages, unchangeable. Our theorizing about the nature of the real is a fallible estimation, the best that can be done at this time, in this particular state of the art. Our science is a historical phenomenon: it is one transitory state of things in an ongoing process.

This limitation cannot, of course, be justifiedly construed as a defect of science. It is an inevitable feature of whatever can be produced by imperfect triangulation from limited experience, and we must not complain of what cannot be helped. It is not a shortcoming of science as compared with other methods of inquiry, because other methods cannot overcome it either. (If they could, so could science.) It is a limitation—an inherent limitation of the enterprise as we humans do and must conduct it. The aim or goal of science is to provide answers; but answers of certified correctness, definitive and final answers, are simply not available. The complete realization of the aims of science is something that will ever remain in the realm of aspiration and not that of achievement.

4. The Problem of Progress

The historical course of natural science is a sequence of radical changes of mind—even about fundamentals. How, then, can science be said to progress? How can we speak of an advance rather than a fortuitous ebb and flow? If science neither provides nor approaches the ultimate truth about nature, how can we say that it has the directionality that is essential to progress, in contradistinction to mere movement?

Historians and analysts of science are often heard to complain about the absence of any clear sense of the direction of development in the structure of modern scientific work. As one recent writer puts it: "The blackest defect in the history of science, the cause of dullest despair for the historian, lies in the virtual absence of any general historical sense of the way science has been working for the last hundred years."[5] Any such lack of direction at once disappears when we turn from the content of scientific discovery to its tools and their mode of employment—in short, when we turn to the technological side of the matter. For all of recent science has a clear thrust of development—using ever more potent instruments to press ever further outward in the exploration of physical parameter-space, forging more and more powerful physical and conceptual instrumentalities for the identification and analysis of new phenomena.

Some recent writers on the philosophy of science, influenced by the doctrines of Thomas Kuhn's book on scientific revolutions,[6] tend to stress the ideational discontinuities produced by innovation in the history of science. Scientific change, they rightly maintain, is not just a matter of marginal revisions of opinion within a fixed and stable framework of concepts; the crucial developments involve a change in the conceptual apparatus itself. When this happens, there is a replacement of the very content of discussion, a shift in what's being talked about that renders successive positions incommensurable. The change from the Newtonian to the Einsteinian concept of time exemplifies a shift of just this sort. These discontinuities of meaning make it impossible (so it is said) to say justifiedly that the latter stages somehow represent a better treatment of the same subject matter, since the very subject has changed.[7] The radical discontinuity of meanings also complicates the idea of scientific progress. For if the later stage of discussion is conceptually disjointed from the earlier, how could one consider the later as an improvement upon the earlier? The replacement of one thing by something else of a totally different sort can hardly qualify as meliorative. (One can improve upon one's car by getting a better car, but one cannot improve it by getting a computer or a dishwashing machine.)

To draw this sort of implication from the meaning-shift thesis is, however, to be overhasty. For real progress is indeed made, although this progress does not proceed along purely theoretical but along practical lines. Once one sees the validation of science as lying ultimately in the sphere of its applications, one also sees that the progress of science must be taken to rest on its pragmatic improvement: the increasing success of its applications in problem solving and control, in cognitive and physical mastery over nature.

The progressiveness of science is most strikingly and decisively manifested on its technological side. Science is marked by an ever-expanding predictive and physical control over nature.[8] The control that lies at the root of progress in science is not something arcane, sophisticated, and heavily theory-laden. It turns on the fact that *any* enhancement in control—any growth of our technological mastery over nature—will have involvements that are also discernible at the *grosso modo* level of our everyday life conceptions and dealings. How our control is extended will generally be a very sophisticated matter, but any fool can see *that* our control has been extended in innumerable ways. Since the days of Bacon and Hobbes, it has been recognized that the conception of the application of science to human ends (*scientia propter potentiam*) provides a perfectly workable basis for taking the expanding horizons of technological capacity as an index of scientific progress.[9]

This technological dimension endows scientific change with a continuity it lacks at the level of its ideas and concepts—a continuity that finds its expression in the persistence of problem-solving tasks in the sphere of praxis. Despite any semantic or ideational incommensurability between a scientific theory and its latter-day replacements, there remains the factor of the pragmatic commensurability that can (by and large) be formulated in suitable extrascientific language.[10]

Later science is better science—that is, better warranted science. And warrant inheres in the application of standards and criteria of substantiation—of scientific methods, in short. The adequacy of these methods and, above all, the superiority of their refinements are attested by considerations of enhanced pragmatic efficacy in application.

Issues of technological superiority are far less sophisticated, but also far more manageable, than issues of theoretical superiority. Dominance in the technological power to produce intended results tends to operate across the board. For the factors determinative of technical superiority in prediction and control operate at a grosser and more rough-and-ready level than those of theoretical content. At the level of praxis, we can operate to a relatively large degree with the lingua franca of everyday affairs and make our comparisons on this basis. Just as the merest novice can detect a false note in the musical performance of a master player whose activities he could not begin to emulate, so the malfunctioning of a missile or computer can be detected by a relative amateur. The superiority of modern over Galenic medicine requires few, if any, subtle distinctions.

Traditional theories of scientific progress join in stressing the capacity of the "improved" theories to accommodate new facts. Agreeing with this emphasis on new facts, we must, however, recognize two distinct routes to this destination: the predictive (via theory) and the productive (via technology). It would appear proper to allow both of these routes to count. To correct the overly theoretical bias of traditional philosophy of science requires a more ample recognition of the role of technology-cum-production. To say this is not, of course, to deny that theory and technology stand in a symbiotic and mutually supportive relationship in scientific inquiry. But the crucial fact is that the effectiveness of the technological instrumentalities of praxis can clearly be assessed without appeal to the cognitive content of the theory brought to bear in their devising.

In maintaining that the successive theses of science represent not just change but progress, one does indeed stand committed to the view that in some fashion or other something that is strictly comparable is being improved upon. But the items at issue need not be the theory-laden substance of scientific claims; they can revolve around the practical concerns of our commonplace affairs. The comparisons need not be made at the level of scientific theorizing but at the rudimentary level of common-sense affairs facilitated by their technological implementation.

At bottom, the progress of science manifests itself most clearly in practical rather than theoretical regards. The progressiveness of science hinges crucially on its applications; it resides in the pragmatic dimension of the enterprise—the increasing success of its applications in problem solving and control, in its affording not only cognitive but physical mastery over nature. A new theory need not explain the purported facts of the old one it replaces, because these "facts" need not remain facts: the new theory may revise or dismiss them. (The phenomenon of "emission of Phlogiston" disappears with the arrival of Lavoisier's oxidation.) But the practical successes of the old theory in enabling us to predict occurrences or to achieve control at the unrefined level of developments describable in the language of everyday life is something relatively unproblematic, and thus relatively secure.

Theorizing is highly contextual with regard to historical and cultural conditions. Science can even change its mind with respect to the very facts themselves, seeing that theory plays a critical role in determining what is to count as a fact. The "natural motion" of a celestial body is one thing for Aristotle and another for Galileo; an evil-eye hexing is one sort of thing for an unlettered peasant and quite another for an urban psychologist. Every theory configuration is itself allowed to determine what sorts of phenomena count as data and how these data are to be construed. There is no theory-neutral body of facts that makes a direct comparison possible; no neutral, science-external, "higher" standpoint to provide a basis of comparison. But the case is very different as regards praxis, and, in particular, praxis at the level of everyday-life proceedings. The lab explodes; the rocket doesn't fly. Here we do indeed have theory-neutral facts for assessing the comparative merits of rival theories. But such comparisons of effective "control over nature" operate at the level of pre- or subscientific dealings and affairs describable in the lingua franca of everyday life, which provides the neutral ground shared in common by different theory postures. Application becomes the touchstone.

From where we stand in the epistemic dispensation, there is no way of attaining a higher vantage point—no God's-eye view for comparison. What we do fortunately have as a common basis across the divides of scientific change is not a higher but rather a lower standpoint: the crude vantage point of ordinary everyday life. No sophisticated complexities are needed to say that one stage in the career of science is superior to another in launching rockets and curing colds and exploding bombs. These applications operate extensively at the level of the ordinary, everyday concepts of natural-language discourse. This level is relatively stable and unchanging precisely because of the crudity of its concerns; its concepts remain relatively fixed throughout the ages and lie deep beneath the changing surface of scientific sophistication.

Its technological and applicative dimension endows science with a theory era-transcending comparability that it lacks at the level of its ideas and concepts. The ancient Greek physician and the modern medical practitioner might talk of the problems of their patients in very different and conceptually incommensurate ways (say, an imbalance in humors to be treated by countervailing changes in diet or regimen versus a bacterial infection to be treated by administering an antibiotic). But at the pragmatic level of practical control—that is, at the level of removing those symptoms of their patients (pain, fever, dizziness, etc.) that are describable in much the same terms in antiquity as today—both are working on "the same problem."

A later scientific theory certainly need not preserve the content of the earlier ones; its descriptive and taxonomic innovations can make for semantical discontinuity. Nor, again, need it preserve the theoretical successes of the earlier theories—their explanatory successes at elaborating interrelations among their own elements (for example, answering questions about the operations of the luminiferous ether). But a later and progressively superior theory must preserve and improve upon the practical successes of its predecessors when these practical issues are formulated in the rough-and-ready terms of everyday-life discourse.

The ultimate arbiter of scientific progress is praxis—and praxis at a level where effectiveness is discernible from the rude and crude posture of ordinary life. And so, notwithstanding its instability and changeability at the level of theoretical claims, science does indeed progress not, to be sure, by way of "approaching the ultimate truth" but by providing us with increasingly powerful instrumentalities for prediction and control. Once due prominence is given to the factor of *control over nature* in the pre- or subtheoretical construction of this idea, the substantiation of imputations of scientific progress becomes a more manageable project than it could ever be on an internal, content-oriented basis.[11]

4 Complexity Escalation as an Obstacle to Completing Science

Synopsis
(1) In a complex world, the natural dynamics of rational inquiry will inevitably exhibit a tropism toward increasing complexity. (2) To be sure, a penchant for simplicity is built into the inductive process of scientific method. But this procedurally methodological commitment does not prejudge or prejudice the substantively ontological issue of the complexity of nature. Our inductive commitment to simplicity is a matter of the procedural principle of least effort—an inseparable aspect of pragmatic rationality. (3) Scientific theorizing is an inductive projection from the available data. But data availability is bound to improve with the changing state of the technological art—engendering a dynamism that ongoingly destabilizes the existing state of science so as to engender greater sophistication. The increasing complexity of our world picture is a striking phenomenon throughout this process. It is so marked, in fact, that natural science has in recent years been virtually disintegrating before our very eyes. (4) And this phenomenon characterizes all of science—the human sciences included. Indeed, complexification and its concomitant destabilization are by no means phenomena confined to the domain of science—they pervade the entire range of our knowledge. (5) Natural science has been growing exponentially throughout recent history in terms of its size and scope as a human enterprise. (6) However, scientific knowledge does not correlate with the brute volume of scientific information, but only with its logarithm. (7) The rationale for this situation lies in the way in which the most significant information is always enveloped by lesser detail, obscured amid a fog of insignificance. (8) In consequence, the progress of knowledge makes ever-escalating demands upon researchers, a circumstance that bears in a profound and ominous way upon the issue of the growth of scientific knowledge over time. (9) To be sure, the phenom-

enon of a logarithmic return on investment in natural science turns on the fact that anything deserving of the name of scientific knowledge must be information of the highest quality. The result of this state of affairs is that the construction of an ever more cumbersome and complex account of things is part of the unavoidable cost of scientific progress.

1. Spencer's Law: The Dynamics of Cognitive Complexity

We naturally adopt throughout rational inquiry—and accordingly throughout natural science—the methodological principle of rational economy to "try the simplest solutions first" and then make this do as long as it can. This means that historically the course of inquiry moves in the direction of ever increasing complexity. The developmental tendency of our intellectual enterprises—natural science among them—is generally in the direction of greater complication and sophistication.

In a complex world, the natural dynamics of the cognitive process exhibit an inherent tropism toward increasing complexity. Herbert Spencer argued long ago that evolution is characterized by von Baer's law of development "from the homogeneous to the heterogeneous" and thereby produces an ever-increasing definition of detail and complexity of articulation.[1] As Spencer saw it, organic species in the course of their development confront a successive series of environmental obstacles, and with each successful turning along the maze of developmental challenges the organism becomes selectively more highly specialized in its bio-design, and thereby more tightly attuned to the particular features of its ecological context.[2]

This view of the developmental process may or may not be correct for biological evolution, but there can be little question about its holding for cognitive evolution. Rational beings will of course try simple things first and thereafter be driven step by step toward an ever enhanced complexification. In the course of rational inquiry we try the simple solutions first, and only thereafter, if and when they cease to work—when they are ruled out by further findings (by some further influx of coordinating information)—do we move on to the more complex. Things go along smoothly until an oversimple solution becomes destabilized by enlarged experience. For a time we get by with the comparatively simpler options—until the expanding information about the world's modus operandi made possible by enhanced new means of observation and experimentation insists otherwise. With the expansion of knowledge those new accessions make ever increasing demands. Thus evolution, be it natural or rational—whether of animal species or of literary genres—ongoingly confronts us with products of greater and greater complexity.[3]

Throughout rational inquiry we seek the simplest, the most economical theory framework that is conveniently available to resolve our explanatory questions. We want, in sum, to find the most economical theory accommodation for the amplest body of currently available experience. Induction—here short for "the scientific method" in general—proceeds by way of constructing the most straightforward and economical structures able to house the available data comfortably while yet affording answers to our questions.[4] Accordingly, economy and simplicity serve as cardinal directives for inductive reasoning, whose procedure is that of the precept: "Resolve your cognitive problems in the simplest, most economical way that is compatible with a sensible exploitation of the information at your disposal." But we always encounter limits here.

Our cognitive efforts manifest a Manichaean-style struggle between complexity and simplicity—between the impetus to comprehensiveness (amplitude) and the impetus to system (economy). We want our theories to be as extensive and all-encompassing as possible and at the same time to be elegant and economical. The first desideratum pulls in one direction, the second in the other. And the accommodation reached here is never actually stable. As our experience expands in the quest for greater adequacy and comprehensiveness, the old theory structures become destabilized—the old theories no longer fit the full range of available fact. And so the theoretician goes back to the old drawing board. What he comes up with here is—and in the circumstances must be—something more elaborate, more complex than what was able to do the job before those new complications arose (although we do, of course, sometimes achieve local simplifications within an overall global complexification). We make do with the simple, but only up to the point when the demands of adequacy force additional complications upon us. An inner tropism toward increasing complexity is thus built into the very nature of the scientific project as we have it.

The same is true also for technological evolution, with cognitive technology emphatically included. Be it in cognitive or in practical matters, the processes and resources of yesteryear are rarely, if ever, up to the demands of the present. In consequence, the life environment we create for ourselves grows increasingly complex. The Occam's Razor injunction, "Never introduce complications unless and until you actually require them," accordingly represents a defining principle of practical reason that is at work within the cognitive project as well. And because we try the simplest solutions first, making simple solutions do until circumstances force one to do otherwise, it transpires that in the development of knowledge—as elsewhere in the domain of human artifice—progress is always a matter of complexification. An inherent impetus toward greater complexity pervades the entire realm of human creative effort. We find it in art; we find it in technology; and we certainly find it in the cognitive domain as well.[5]

2. The Principle of Least Effort and the Methodological Status of Simplicity-Preference in Science

An eminent philosopher of science has maintained that "in cases of inductive simplicity it is not economy which determines our choice.... We make the assumption that the simplest theory furnishes the best predictions. This assumption cannot be justified by convenience; it has a truth character and demands a justification within the theory of probability and induction."[6] This perspective is gravely misleading. What sort of consideration could possibly justify the supposition that "the simplest theory furnishes the best prediction"? Any such belief is surely unwarranted and inappropriate. There is simply no cogent rationale for firm confidence in the simplicity of nature. To claim the ontological simplicity of the real is somewhere between hyperbolic and absurd.

The matter becomes far less problematic, however, once one approaches it from a methodological rather than a substantive point of view, for considerations of rational economy and convenience of operation obviously militate for inductive systematicity. Seeing that the simplest answer is (*eo ipso*) the most economical one to work with, rationality creates a natural pressure toward economy—toward simplicity insofar as other things are equal. Our eminent theorist has things upside down here: it is in fact methodology that is at issue rather than any factual presumption.

Suppose, for example, that we are asked to supply the next member of a series of the format 1, 2, 3, 4.... We shall straightaway respond with 5, supposing the series to be simply that of the integers. Of course, the actual series might well be 1, 2, 3, 4, 11, 12, 13, 14, 101, 102, 103, 104 ..., with the correct answer thus eventuating as 11 rather than 5. But while we cannot rule such possibilities out, they do not for an instant deter our inductive proceedings. For the inductively appropriate course lies with the production rule that is the simplest issue-resolving answer—the simplest resolution that meets the conditions of the problem. And we take this line not because we know a priori that this simplest resolution will prove to be correct. (We know no such thing!) Rather we adopt this answer, provisionally at least, just exactly because it is the least cumbersome and most economical way of providing a resolution that does justice to the facts and demands of the situation. We recognize that other possibilities of resolution exist but ignore them until further notice, exactly because there is no cogent reason for giving them favorable treatment *at this stage*. (After all, once we leave the safe harbor of simplicity behind there are always multiple possibilities for complexification, and we lack any guidance in moving one way rather than another.)

Throughout inductive inquiry in general, and scientific inquiry in particular, we seek to provide a descriptive and explanatory account that provides the simplest, least complex way of accommodating the data that experience (experimentation and observation) has put at our disposal. When something simple accomplishes the cognitive tasks in hand, as well as some more complex alternative, it is foolish to adopt the latter. After all, we need not presuppose that the world somehow is systematic (simple, uniform, and the like) to validate our penchant for the systematicity of our cognitive commitments.

Henri Poincaré has remarked that:

> [Even] those who do not believe that natural laws must be simple, are still often obliged to act as if they did believe it. They cannot entirely dispense with this necessity without making all generalization, and therefore all science, impossible. It is clear that any fact can be generalised in an infinite number of ways, and it is a question of choice. The choice can only be guided by considerations of simplicity. . . . To sum up, in most cases every law is held to be simple until the contrary is proved.[7]

These observations are wholly in the right spirit. As cognitive possibilities proliferate in the course of theory-building inquiry, a principle of choice and selection becomes requisite. And here economy—along with its other systematic congeners, simplicity, uniformity, and the like—are the natural guideposts. We subscribe to the inductive presumption in favor of simplicity, uniformity, normality,[8] and so on, not because we know or believe that matters always stand on a basis that is simple, uniform, normal, etc.—surely we know no such thing!—but because it is on this basis alone that we can conduct our cognitive business in the most advantageous, the most economical way. In scientific induction we exploit the information at hand so as to answer our questions in the most straightforward, the most economical way.

Throughout inductive situations, we are invariably called on to answer questions whose resolution is beyond the secure reach of information at hand. To accomplish this we contrive our problem resolutions along the lines of least resistance, seeking to economize our cognitive effort by using the most direct workable means to our ends. Whenever possible, we analogize the present case to other similar ones, because the introduction of new patterns complicates our cognitive repertoire. And we use the least cumbersome viable formulations because they are easier to remember and more convenient to use. The rationale of the other-things-equal preferability of simpler solutions over more complex ones is obvious enough. Simpler solutions are less cumbersome to store, easier to take hold of, and less difficult to work with. It is indeed economy and convenience that determine our regulative predilection for simplicity and systematicity in general. Our prime motivation is to get by with a minimum of complication, to adopt strategies of question resolution that enable us among other things: (1) to continue with existing solutions unless and until the epistemic circumstances compel us to introduce changes (uniformity), (2) to make do with the same processes insofar as possible (generality), and (3) to keep to the simplest process that will do the job (simplicity). Such a perspective combines the commonsensical precept, "Try the simplest thing first," with a principle of burden of proof: "Maintain your cognitive commitments until there is good reason to abandon them."

When other things are anything like equal, simpler theories are bound to be operationally more advantageous. We avoid needless complications whenever possible, because this is the course of an economy of effort. It is the general practice in scientific theory construction to give preference to

- one-dimensional rather than multidimensional modes of description,
- quantitative rather than qualitative characterizations,
- lower- rather than higher-order polynomials,
- linear rather than nonlinear differential equations.

The comparatively simpler is for this very reason easier to work with. In sum, we favor uniformity, analogy, simplicity, and the like because they ease our cognitive labor. On such a perspective, simplicity is a concept of the practical order, pivoting on being more economical to use—that is, less demanding of resources. The key principle is that of the rational economy of means for the realization of given cognitive ends, of getting the most effective answer we can with the least complication. Complexities cannot be ruled out, but they must always pay their way in terms of increased systemic adequacy! It is thus methodology and not metaphysics that grounds our commitment to simplicity and systematicity.

The rational economy of process is the crux here. This methodological commitment to rational process does not prejudge or prejudice the substantively ontological issue of the complexity of nature. Natural science is emphatically not bound to a Principle of Simplicity in Nature. There really are no adequate grounds for supposing the simplicity of the world's makeup. Instead, the so-called Principle of Simplicity is really a principle of complexity management: "Feel free to introduce complexity in your efforts to describe and explain nature's ways, but only when and where it is really needed. Insofar as possible, keep it simple! Only introduce as much complexity as you really need for your scientific purposes of description, explanation, prediction, and control." Such an approach is eminently sensible. But of course such a principle is no more than a methodological rule of procedure for managing our cognitive affairs. Nothing entitles us to transmute this methodological precept into a descriptive/ontological claim to the effect that nature is simple—let alone of finite complexity.

Our systematizing procedures in science pivot on the injunction always to adopt the most economical (simple, general, straightforward) solution that meets the demands of the situation. This penchant for inductive systematicity reflected in the conduct of inquiry is simply a matter of striving for rational economy. It is based on methodological considerations that are governed by an analogue of Occam's Razor—a principle of parsimony to the effect that needless complexity is to be avoided—*complicationes non multiplicandae sunt praeter necessitatem*. Given that the inductive method, viewed in its practical and methodological aspect, aims at the most efficient and effective means of question resolution, it is only natural that our inductive precepts should direct us always to begin with the most systematic, and thereby economical, device that can actually do the job at hand.[9]

It clearly makes eminent sense to move onward from the simplest (least complex) available solution to introduce further complexities when and as—but only when and as—they are forced upon us. Simpler (more systematic) answers are more easily codified, taught, learned, used, investigated, and so on. The regulative principles of convenience and economy in learning and inquiry suffice to provide a rational basis for systematicity preference. Our preference for simplicity, uniformity, and systematicity in general is now not a matter of a substantive theory regarding the nature of the world, but one of search strategy—of cognitive methodology. In sum, we opt for simplicity (and systematicity in general) in inquiry not because it is truth-indicative, but because it is teleologically more effective in conducing to the efficient realization of the goals of inquiry. We look for the dropped coin in the lightest spots nearby, not because this is—in the circumstances—the most probable location but because it represents the most sensible strategy of search: if it is not there, then we just cannot find it at all.

On such a view, inductive systematicity with its penchant for simplicity comes to be seen as an aspect, not of reality as such, but of our procedures for its conceptualization and accordingly of our conception of it, or, to be more precise, of our manner of conceptualizing it. Simplicity preference (for example) is based on the strictly method-oriented practical consideration that the simple hypotheses are the most convenient and advantageous for us to put to use in the context of our purposes. There is thus no recourse to a substantive (or descriptively constitutive) postulate of the simplicity of nature; it suffices to have recourse to a regulative (or practical) precept of economy of means. And in its turn, the pursuit of cognitive systematicity is ontologically neutral: it is a matter of conducting our question-resolving endeavors with the greatest economy. The Principle of Least Effort is in control here—the process is one of maximally economic means to the attainment of chosen ends. This amounts to a theoretical defense of inductive systematicity that in fact rests on practical considerations relating to the efficiencies of method. Accordingly, inductive systematicity is best approached with reference, not to reality as such—or even merely our conception of it—but to the ways and means we employ in conceptualizing it. It is noncommittal on matters of substance, representing no more than a determination to conduct our question-resolving endeavors with the greatest economy. For in inquiry, as elsewhere, rationality enjoins us to employ the maximally economic means to the attainment of chosen ends.

It is thus important to distinguish between substantive and methodological considerations and to separate economy of means from substantive economy. For process is one thing and product another. Simple tools or methods can, suitably used, create complicated results. A simple cognitive method, such as trial and error, can ultimately yield complex answers to difficult questions. Conversely, simple results are sometimes brought about in complicated ways; complicated methods of inquiry or problem solving might arrive at simple and uncomplicated solutions. Our commitment to simplicity in scientific inquiry accordingly does not, in the end, prevent us from discovering whatever complexities are actually there. And the commitment to inductive systematicity in our account of the world remains a methodological desideratum regardless of how complex or untidy that world may ultimately turn out to be.

3. Complexification and the Disintegration of Science

It is worthwhile to examine somewhat more closely the ramifications of complexity in the domain of cognition, now focusing upon science in particular. Progress in natural science is a matter of dialogue or debate in a reciprocal interaction between theoreticians and experimentalists. The experimentalists probe nature to discern its reactions, to seek out phenomena. The theoreticians take the resultant data and weave about them a fabric of hypotheses that is able to resolve our questions. Seeking to devise a framework of rational understanding, they construct their explanatory models to accommodate the findings that the experimentalists put at their disposal. Thereafter, once the theoreticians have had their say, the ball returns to the experimentalists' court. Employing new, more powerful means for probing nature, they bring new phenomena to view, new data for accommodation. Precisely because these data are new and inherently unpredictable on the basis of earlier knowledge, they often fail to fit the old theories. Theory extrapolations from the old data could not encompass them; the old theories do not accommodate them. A disequilibrium thus arises between available theory and novel data, and at this stage, the ball reenters the theoreticians' court. New theories must be devised to accommodate the new, nonconforming data. Accordingly, the theoreticians set about weaving a new theoretical structure to accommodate the new data. They endeavor to restore the equilibrium between theory and data once more. When they succeed, the ball returns to the experimentalists' court, and the whole process starts over again.

Scientific theory formation is, in general, a matter of spotting a local regularity of phenomena in parametric space and then projecting it across the board, maintaining it globally. But the theoretical claims of science are themselves never small-scale and local—they are not spatiotemporally localized and they are not parametrically localized either. They stipulate—quite ambitiously—how things are always and everywhere. But with the enhancement of investigative technology, the window through which we can look out upon nature's parametric space becomes constantly enlarged. In developing natural science we use this window of capability to scrutinize parametric space, continually augmenting our database and then generalizing upon what we see. What we have here is not a lunar landscape where once we have seen one sector we have seen it all, and where theory projections from lesser data generally remain in place when further data comes our way. Instead it does not require a sophisticated knowledge of history of science to realize that our worst fears are usually realized—that our theories seldom if ever survive intact in the wake of substantial extensions in our cognitive access to new sectors of the range of nature's phenomena. The history of science is a sequence of episodes of leaping to the wrong conclusions because new observational findings indicate matters are not quite so simple as heretofore thought. As ample experience indicates, our ideas about nature are subject to constant and often radical, change-demanding stresses as we "explore" parametric space more extensively. The technologically mediated entry into new regions of parameter space constantly destabilizes the attained equilibrium between data and theory. Physical nature can exhibit a very different aspect when viewed from the vantage point of different levels of sophistication in the technology of nature-investigator interaction. The possibility of change is ever present. The ongoing destabilization of scientific theories is the price we pay for operating a simplicity-geared cognitive methodology in an actually complex world.

The methodology of science thus embodies an inherent dialectic that moves steadily from the simpler to the more complex, and the developmental route of technology sails on the same course. We are driven in the direction of ever greater complexity by the principle that the potential of the simple is soon exhausted and that high capacity demands more elaborate and powerful processes and procedures. The simpler procedures of the past are but rarely adequate to the needs of the present—had they been so today's questions would have been resolved long ago and the issues at stake would not have survived to figure on the present agenda. Scientific progress inherently involves an inexorable tendency to complexification in both its cognitive and its ideational dimension. What we discover in investigating nature always must in some degree reflect the character of our technology of observation. It is always something that depends on the mechanisms with which we search.[10]

Induction with respect to the history of science itself—a constant series of errors of oversimplification—soon undermines our confidence that nature operates in the way we would deem the simpler. On the contrary, the history of science is an endlessly repetitive story of simple theories giving way to more complicated and sophisticated ones. The Greeks had four elements; in the nineteenth century Mendeleev had some sixty; by the 1900s this had gone to eighty, and nowadays we have a vast series of elemental stability states. Aristotle's cosmos had only spheres; Ptolemy's added epicycles; ours has a virtually endless proliferation of complex orbits that only supercomputers can approximate. Greek science was contained on a single shelf of books; that of the Newtonian age required a roomful; ours requires vast storage structures filled not only with books and journals but with photographs, tapes, floppy disks, and so on. Of the quantities currently recognized as the fundamental constants of physics, only one was contemplated in Newton's physics: the universal gravitational constant. A second was added in the nineteenth century, Avogadro's constant. The remaining six are all creatures of twentieth century physics: the speed of light (the velocity of electromagnetic radiation in free space), the elementary charge, the rest mass of the electron, the rest mass of the proton, Planck's constant, and Boltzmann's constant.[11]

It would be naive—and quite wrong—to think that the course of scientific progress is one of increasing simplicity. The very reverse is the case: scientific progress is a matter of complexification because oversimple theories invariably prove untenable in a complex world. The natural dialectic of scientific inquiry ongoingly impels us into ever deeper levels of sophistication.[12] In this regard our commitment to simplicity and systematicity, although methodologically necessary, is ontologically unavailing. More sophisticated searches invariably engender changes of mind moving in the direction of an ever more complex picture of the world. Our methodological commitment to simplicity should not and does not preclude the substantive discovery of complexity.

The explosive growth of information of itself countervails against its exploitation for the sake of knowledge enhancement. The problem of coping with the proliferation of printed material affords a striking example of this phenomenon. One is forced to ever higher levels of aggregation, compression, and abstraction. In seeking for the needle in the haystack we must push our search processes to ever greater depths.

This ongoing refinement in the division of cognitive labor that an information explosion necessitates issues in a literal dis-integration of knowledge. The "progress of knowledge" is marked by an ever continuing proliferation of ever more restructured specialties marked by the unavoidable circumstance that any given specialty cell cannot know exactly what is going on even next door—let alone at the significant remove. Our understanding of matters outside our immediate bailiwick is bound to become superficial. At home base one knows the details, nearby one has an understanding of generalities, but at a greater remove one can be no more than an informed amateur.

This disintegration of knowledge is also manifolded vividly in the fact that our cognitive taxonomies are bursting at the seams. Consider the example of the taxonomic structure of physics. In the eleventh (1911) edition of the *Encyclopedia Britannica*, physics is described as a discipline composed of nine constituent branches (e.g., "Acoustics" or "Electricity and Magnetism") which were themselves partitioned into twenty further specialties (e.g., "Thermo-electricity" or "Celestial Mechanics"). The fifteenth (1974) version of the *Britannica* divides physics into twelve branches whose subfields are—seemingly—too numerous for listing. (However, the fourteenth, 1960s edition carried a special article entitled "Physics, Articles on" that surveyed more than 130 special topics in the field.) When the National Science Foundation launched its inventory of physical specialties with the National Register of Scientific and Technical Personnel in 1954, it divided physics into twelve areas with ninety specialties. By 1970 these figures had increased to sixteen and 210, respectively, and the process continues unabated, to the point where people are increasingly reluctant to embark on this classifying project at all.

Substantially the same story can be told for every field of science. The emergence of new disciplines, branches, and specialties is manifest everywhere. As though to negate this tendency and maintain unity, one finds an ongoing evolution of interdisciplinary syntheses—physical chemistry, astrophysics, biochemistry, and so on. The very attempt to counteract fragmentation produces new fragments. Indeed, the phenomenology of this domain is nowadays so complex that some writers urge that the idea of a "natural taxonomy of science" must be abandoned altogether.[13] The expansion of the scientific literature is in fact such that natural science has in recent years been disintegrating before our very eyes. An ever larger number of ever more refined specialties has made it ever more difficult for experts in a given branch of science to achieve a thorough understanding about what is going on even in the specialty next door.

It is, of course, possible that the development of physics may eventually carry us to theoretical unification such that everything that we class among the laws of nature belongs to one grand unified theory—one all-encompassing deductive systematization integrated even more tightly than Newton's *Principia Mathematica*.[14] But the covers of this elegantly contrived book of nature will have to encompass a mass of ever more elaborate diversity and variety. Like a tricky mathematical series, it will have to generate ever more dissimilar constituents that, despite their abstract linkage, are concretely as different as can be. The integration at issue, as the principle of a pyramid, will cover further down an endlessly expansive range and encompass the most variegated components. It will be an abstract unity uniting a concrete mishmash of incredible variety and diversity. The unity of science to which many theorists aspire may indeed come to be realized at the level of concepts and theories shared between different sciences—that is, at the level of ideational overlaps. But for every conceptual commonality and shared element there will emerge a dozen differentiations. The increasing complexity of our world picture is a striking phenomenon throughout the development of modern science.

The lesson of such considerations is clear. Scientific knowledge grows not just in extent but also in complexity, so that science presents us with a scene of ever-increasing complexity. It is thus fair to say that modern science confronts us with a cognitive manifold that involves an ever more extensive specialization and division of labor. The years of apprenticeship that separate master from novice grow ever greater. A science that moves continually from an oversimple picture of the world to one that is more complex calls for ever more elaborate processes for its effective cultivation. As the scientific enterprise itself grows more extensive, the greater elaborateness of its productions requires an ever more intricate intellectual structure for its accommodation. The complexifications of scientific process and product escalate hand in hand. And the process of complexity amplification that Charles S. Peirce took to be revealed in nature through science is unquestionably manifested in the cognitive domain of scientific inquiry itself.

4. The Expansion of Science

The increasing complexity of the scientific enterprise itself is reflected in the fact that research and development expenditures in the United States grew exponentially after World War II, increasing at a rate of some 10 percent per annum. By the mid-1960s, America was spending more on scientific research and development than the entire federal budget before Pearl Harbor. This growth in the costs of science has various significant ramifications and manifestations.

Take manpower, for example, where the recent growth of the scientific community is a particularly striking phenomenon. During most of the present century the number of American scientists has been increasing at 6 percent to yield an exponential growth rate with a doubling time of roughly twelve years.[15] A startling consideration—one often but deservedly repeated—is that well over 80 percent of ever-existing scientists (in even the oldest specialties such as mathematics, physics, and medicine) are alive and active nowadays.[16]

Again, consider the growth of the scientific literature. It is by now a familiar fact that scientific information has been growing at a (reasonably constant) exponential rate over the past several centuries. Overall, the printed literature of science has been increasing at an average of some 5 percent annually throughout the last two centuries, to yield an exponential growth rate with a doubling time of circa fifteen years—an order-of-magnitude increase roughly every half-century. The result is a veritable flood of scientific literature. The *Physical Review* is currently divided into six parts, each of which is larger than the whole journal was a decade or so ago. It is reliably estimated that, from the start, about 10 million scientific papers have been published and that currently some 30,000 journals publish some 600,000 new papers each year. In fact, it is readily documented that the number of books, of journals, and of journal papers has been increasing at an exponential rate over the recent period.[17] By 1960, some 300,000 different book titles were being published in the world, and the two decades from 1955 to 1975 saw the doubling of titles published in Europe from around 130,000 to over 270,000.[18] Science has had its full share of this literature explosion. The amount of scientific material in print is of a scope that puts it beyond the reach not only of individuals but also of institutions as well. No university or institute has a library vast enough to absorb or a faculty large enough to digest the relevant products of the world's printing presses.

Then too there is the massively increasing budget of science. The historic situation regarding the costs of American science was first delineated in the findings of Raymond Ewell in the 1950s.[19] His study of research and development expenditures in the United States showed that growth here has also been exponential; from 1776 to 1954 we spent close to $40 billion, and half of that was spent after 1948.[20] Projected at this rate, Ewell saw the total as amounting to what he viewed as an astronomical $6.5 billion by 1965—a figure that actually turned out to be far too conservative.

Moreover, the proliferation of scientific facilities has proceeded at an impressive pace over the past hundred years. (In the early 1870s there were only eleven physics laboratories in the British Isles; by the mid-1930s there were more than three hundred;[21] today there are several thousand.) Of course, the scale of activities in these laboratories has also expanded vastly.

It is perhaps unnecessary to dwell at length on the immense cost in resources of the research equipment of contemporary science.[22] Radiotelescopic observatories, low-temperature physics, research hospitals, and lunar geology all involve outlays of a scale that require the funding and support of national governments—sometimes even consortia thereof. In a prophetic vein, Alvin M. Weinberg (then director of the Oak Ridge National Laboratory) wrote: "When history looks at the 20th century, she will see science and technology as its theme; she will find in the monuments of Big Science—the huge rockets, the high-energy accelerators, the high-flux research reactors—symbols of our time just as surely as she finds in Notre Dame a symbol of the Middle Ages."[23] Of course, exponential growth cannot continue indefinitely. The fact remains, however, that science has become an enormous industry with a far-flung network of training centers (schools, colleges, universities), and of production centers (laboratories and research institutes).

Science, in sum, has become a vast and expensive business. But what sort of relationship obtains between resource investment and returns here? Just how productive is the science enterprise?

5. The Law of Logarithmic Returns

There is no question but that the complexification *of* science produced by its growth as a productive enterprise also makes for a complexification *in* science as a cognitive resource. This occurs at a far lesser rate, and it is instructive to consider the reason why.

En route to knowledge we must begin with information. How is one to measure the volume of information generated in a field of scientific or scholarly inquiry? Direct methods are hard to come by, but various oblique ways suggest themselves. The size of the literature of a field (as measured by the sheer bulk of publication in it) affords one possible measure, and there are various other possibilities as well. One can make one's estimates here by way of inputs rather than outputs, measuring, for example, the aggregate time that investigators devote to their topics, or the volume of resource investment in research-supportive technology. So much for the quantitative assessment of *information*. But what about *knowledge*?

While information is a matter of data, knowledge, by contrast, is something more select, more deeply issue-resolving—to wit, *significant and well-consolidated* information. Anyone who has ever struggled with the statistical analysis of masses of data knows viscerally the difference between basic information and the sort of insight at issue with actual knowledge. (Not every insignificant smidgeon of information constitutes knowledge, and the person whose body of information consists of utter trivia really *knows* virtually nothing.)

For the sake of a crude illustration of the information/knowledge distinction, imagine an object-descriptive color taxonomy—for the sake of example, an oversimple one based merely on blue, red, and other. Then that single item of knowledge represented by "knowing the color" of an object—viz., that it is red—is bound up with many different items of (correct) information on the subject (that it is not blue, is rather similar to some shades of other, etc.). As such information proliferates, we confront a situation of redundancy and diminished productiveness. Any knowable fact is always surrounded by a vast penumbral cloud of relevant information. As our information grows to be ever more extensive, those really significant facts become more difficult to discern.

Knowledge certainly increases with information, but at a far less than proportional rate.[24]

A single step in the advancement of knowledge is thus accompanied by a massive increase in the proliferation of information. Thus with n concepts, one can construct n^2 two-concept combinations, and with m facts, one can project m^2 fact-connecting juxtapositions, each representing some sort of characteristic relationship.[25] Extending the previous example, let us also contemplate shapes in addition to colors, again supposing only three of them: rectangular, circular, other. When we thus combine color and shape there will be $9 = 3 \times 3$ possibilities in the resultant (cross-) classification. So with this enlarged, dual-aspect piece of knowledge (color + shape) we also launch into a vastly amplified—specifically multiplied—information spectrum over that increased classification space. In moving from n to $n + 1$ cognitive parameters, knowledge is increased additively but the information field expands multiplicatively.

It is instructive to view the matter from a different vantage point. Knowledge commonly develops via distinctions (A versus non-A) that are introduced with ever greater elaboration and sophistication to address the problems and difficulties that one encounters with less sophisticated approaches. A situation obtains that is analogous to the game of Twenty Questions with an exponentially exfoliating possibility-space unfolding stepwise $(2, 4, 8, 16, \ldots, 2^n, \ldots)$. With n descriptors one can specify for 2^n potential descriptions that specify exactly how, over all, a given object may be characterized. When we add a new descriptor we increase by one additional unit the amount of knowledge but double the amount of available information. The information at hand grows with 2^n, but the knowledge acquired merely with n. With conceptual elaboration increases in informative detail always ratchet up exponentially. The cognitive yield of increasing information is subject to dramatically diminishing returns.

Consider another example, this time from the field of paleography. The yield of knowledge from information afforded by legibility-impaired manuscripts, papyri, and inscriptions in classical paleography provides an instructive instance. If one can decipher 70 percent of the letters in such a manuscript one can reconstruct a given phrase. If one can make out 70 percent of the phases one can pretty well figure out the sentences. If one can read 70 percent of its sentences one can understand the message of the whole text. So in favorable conditions some one-third of the letters suffice to carry the whole message. A vast load of information stands coordinate with a modest body of knowledge. From the standpoint of knowledge, information is highly redundant, albeit unavoidably so. A helpful perspective on this situation comes to view through the communication-theoretic conception of "noise," seeing that expanding bodies of information encompass so much cognition-impeding redundancy and unhelpful irrelevancy that it takes successive many-fold increases in information to effect successive fixed-size increases in actual knowledge.

Two ideas can profitably be brought together in this context:

(1) Knowledge is distinguished from mere information as such by its significance. In fact: *Knowledge is simply particularly significant information*—information whose significance exceeds some threshold level (say *q*). (In principle there is room for variation here as one sets the quality level of entry qualification and the domain higher or lower.)

(2) The significance of *additional* information is determined by its impact upon *pre-existing* information. Importance or significance in this sense is a matter of the relative (percentage-wise) increase that the new acquisitions effect upon the body of pre-existing information (I), which may—to reiterate—be estimated in the first instance by the sheer volume of the relevant body of information: the literature of the subject, as it were. Accordingly: *The significance of incrementally new information can be measured by the ratio of the increment of new information to the volume of information already in hand*: $\Delta I / I$.

Putting these two considerations together, we find that a new item of actual knowledge is one for which the ratio of information increments to preexisting information exceeds a fixed threshold-indicative quantity *q*:

$$\frac{\Delta I}{I} \geq q$$

Thus knowledge-constituting *significant* information arises when the proportional extent of the change effected by a new item in the preexisting situation (independent of what that preexisting situation is) exceeds a duly fixed threshold.

On the basis of such a construal of *knowledge*, it follows that the cumulative total amount of knowledge (*K*) encompassed in an overall body of information of size *I* is given by the *logarithm* of *I*. This is so because we have

$$K = \int \frac{dI}{I} = \log I + \text{const} = \log cI$$

where *K* represents the volume of actual *knowledge* that can be extracted from a body of bare *information I*.[26] (Here the constant at issue is open to treatment as a unit-determinative parameter of the measuring scale, so that the equation at issue can be simplified without loss to $K = \log I$.)[27] We accordingly arrive at the Law of Logarithmic Returns governing the extraction of significant knowledge from bodies of mere information.

The ramifications of such a principle for cognitive progress are not difficult to discern. Nature imposes increasing resistance barriers to intellectual as well as to physical penetration. Consider the analogy of extracting air for creating a vacuum. The first 90 percent comes out rather easily. The next 9 percent is effectively as difficult to extract as all that went before. The next 0.9 percent is just as difficult. And so on. Each successive order-of-magnitude step involves a massive cost for lesser progress; each successive fixed-size investment of effort yields a substantially diminished return. Intellectual progress is exactly the same: when we extract knowledge (i.e., high-grade, nature-descriptively significant information) from mere information of the routine, common, garden variety, the same sort of quantity/quality relationship obtains. Initially a sizable proportion of the available is high grade—but as we press further this proportion of what is cognitively significant gets ever smaller. To double knowledge we must quadruple information.

It should thus come as no surprise that knowledge coordinates with information in multiplicative layers. With texts we have the familiar stratification: article/chapter, book, library, library system. Or pictographically: sign, scene (= ordered collection of signs), cartoon (= ordered collection of scenes to make it a story). In such layering, we have successive levels of complexity corresponding to successive levels of informational combining combinations, proportional with n, n^2, n^3, etc. The logarithm of the levels—$\log n, 2\log n, 3\log n$, etc.—reflects the amount of "knowledge" that is available through the information we obtain about the state of affairs prevailing at each level. This increases only one unit step at a time, despite the exponential increase in information.

The general purport of such a Law of Logarithmic Returns as regards expanding information is clear enough. Letting $K(I)$ represent the quantity of knowledge inherent in a body of information I, we begin with our fundamental relationship: $K(I) \approx \log I$. On the basis of this fundamental relationship, the knowledge of a two-sector domain increases additively notwithstanding a multiplicative explosion in the amount of information that is now upon the scene. For if the field (F) under consideration consists of two subfields (F_1 and F_2), then because of the cross-connections obtaining within the information domains at issue, the overall information complex will take on a multiplicative size:

$$I = \inf(F) = \inf(F_1) \times \inf(F_2) = I_1 \times I_2$$

With compilation, information is multiplied. But in view of the indicated logarithmic relationship, the knowledge associated with the body of compound information I will stand at:

$$K(I) = \log I = \log(I_1 \times I_2) = \log I_1 + \log I_2 = K(I_1) + K(I_2)$$

The knowledge obtained by joining two information domains (subfields) into an overall aggregate will (as one would expect) consist simply of adding the two bodies of knowledge at issue. Whereas compilation increases information by multiplicative leaps and bounds, the increase in knowledge is merely additive.

6. The Rationale and Implication of the Law

The Law of Logarithmic Returns constitutes an epistemological analogue of the old Weber-Fechner law of psychophysics, which asserts that inputs of geometrically increasing magnitude are required to yield perceptual outputs of arithmetically increasing intensity. This makes for a parallelism of perception and conception in this regard: on both sides we have it that informational inputs of geometrically increasing magnitude are needed to provide for cognitive outputs of arithmetically increasing magnitude. This is illustrated in table 4.1. There is an immense K/I imbalance: with ongoingly increasing information I, the corresponding increase in knowledge K shrinks markedly. To increase knowledge by equal steps we must amplify information by successive orders of magnitude.

Table 4.1. The Structure of the Knowledge/Information Relation.

I	Ic (with $c = .1$)	$\log cI$ ($=K$)	$\Delta(K)$	K as % of I	$\Delta(K)$ as % of K
100	10	1	—	1.0	—
1,000	100	2	1	0.2	50
10,000	1000	3	1	0.03	33
100,000	10,000	4	1	0.004	25

Perhaps the principal reason for such a K/I imbalance may be found in the efficiency of intelligence in securing a view of the modus operandi of a world whose law structure is comparatively simple, since here one can learn a disproportionate amount of general fact from a modest amount of information. (Note that whenever an infinite series of 0's and 1's, as per 01010101..., is generated—as this series indeed is—by a relatively *simple* law, then this circumstance can be gleaned from a comparatively short initial segment of this series.) We thus have a plausible reason for that K/I disparity: where order exists in the world, intelligence is rather efficient in finding it. (And it is worth observing in this context that if the world were not orderly—were not amenable to the probes of intelligence—then intelligent beings would not and could not have emerged in it through evolutionary mechanisms.)

The matter can be viewed in another perspective. Nature imposes increasing resistance barriers to intellectual as well as to physical penetration. Extracting knowledge from information thus requires ever greater effort. For the greater a body of information, the larger the *patterns of order* that can potentially obtain and the greater the effort needed to bring particular orderings to light. With two-place combinations of the letters *A* and *B* (yielding the four pairs AA, AB, BA, and BB) we have only two possible patterns of order—namely "The same letter all the way through" (AA and BB), and "Alternating letters" (AB and BA). As we add more letters, the possibilities proliferate massively. There is, accordingly, a law of diminishing returns in operation here: each successive fixed-size investment of effort yields a substantially diminished return. Intellectual progress is exactly the same: when we extract actual knowledge (i.e., high-grade, nature-descriptively significant information) from mere information of the routine, common garden variety, the same sort of quantity/quality relationship obtains. Initially a sizable proportion of the available is high grade—but as we press further this proportion of what is cognitively significant gets ever smaller. And so, as science progress, the important discoveries that represent real increases in knowledge are surrounded by an ever increasing penumbra of mere items of information. (The mathematical literature of the day yields an annual crop of over 200,000 new theorems.)[28]

This resultant situation is reflected in Max Planck's appraisal of the demands of scientific progress. He wrote that "*with every advance [in science] the difficulty of the task is increased; ever larger demands are made on the achievements of researchers*, and the need for a suitable division of labor becomes more pressing."[29] The Law of Logarithmic Returns at once characterizes and explains this phenomenon of substantial findings being thicker on the informational ground in the earlier phase of a new discipline and growing ever attenuated in the natural course of progress. The upshot is that increasing complexity is the unavailable accompaniment of scientific progress as ever more elaborate processes are required to engender an equimeritious product.

7. The Growth of Knowledge

The Law of Logarithmic Returns clearly has substantial implications for the rate of scientific progress.[30] In particular, it stipulates a swift and steady decline in the comparative cognitive yield of accession to our body of mere information. With the enhancement of scientific technology, the size and complexity of this body of data inevitably grows, expanding on quantity and diversifying in kind. Historical experience indicates that our knowledge is ongoingly enhanced through this broadening of our data base. Of course, this expansion of knowledge proceeds at a far slower rate than does the increase of bare information.

It is illuminating to look at the implications of this state of affairs in a historical perspective. The salient point is that the growth of knowledge over time involves ever-escalating demands. Progress is always possible—there are no absolute limits. However, increments of the same cognitive magnitude have to be obtained at an ever increased price in point of information development and thus of resource commitment as well. The circumstance that knowledge stands proportionate to the logarithm of information ($K \approx \log I$) means the growth of knowledge over time stands to the increase of information in a proportion fixed by the *inverse* of the volume of already available information:

$$\frac{d}{dt} K \approx \frac{d}{dt} \log I \approx \frac{1}{I} \frac{d}{dt} I$$

The more knowledge we already have in hand, the slower (by a very rapid decline) will be the rate at which knowledge grows with newly acquired information. The larger the body of information we have, the smaller will be the proportion of this information that represents real knowledge.

In developmental perspective, there is good reason to suppose that our body of bare information increases more or less in proportion with our resource investment in information gathering. Accordingly, if this investment grows exponentially over time (as we have seen above to have historically been the case in the recent period), we shall in consequence have it that

$$I(t) \approx c^t \text{ and correspondingly also } \frac{d}{dt} I(t) \approx c^t$$

This means that

$$K(t) \approx \log I(t) \approx \log c^t = t$$

and consequently

$$\frac{d}{dt} K(t) = \text{constant}$$

It will follow on this basis that, since *exponential* growth in I is coordinated with a merely *linear* growth in K, the rate of progress of science in the information-exploding past has actually remained essentially constant.

In fact, it is not all that difficult to find empirical substantiation of our law of logarithmic returns ($K \approx \log I$). The phenomenology of the situation is that while scientific information has increased exponentially in the past (as shown by the exponential increase in journals, scientists, and outlays for the instrumentalities of research)—as was documented above—a good case can be made for maintaining that the progress of authentic scientific knowledge as measured in the sort of first-rate discoveries of the highest level of significance has to all appearances progressed at a more or less linear rate. There is much evidence to substantiate this. It suffices to consider the size of encyclopedias and synoptic textbooks, or again, the number of awards given for "really big contributions"

(Nobel prizes, academy memberships, honorary degrees), or the expansion of the classificatory taxonomy of branches of science and problem areas of inquiry.[31] All of these measures of preeminently cognitive contribution conjoin to indicate that there has in fact been a comparative constancy in year-to-year progress, the exponential growth of the scientific enterprise notwithstanding.

Thus, *viewing science as a cognitive discipline*—a growing body of knowledge whose task is the unfolding of a rational account of the modus operandi of nature—the progress of science stands correlative with its accession of really major discoveries: the seismically significant, cartography-revising insights into nature. This perspective lays the foundation for our present analysis—that the historical situation has been one of a constant progress of science as a cognitive discipline, notwithstanding its exponential growth as a productive enterprise (as measured in terms of resources, money, manpower, publications, etc.). The Law of Logarithmic Returns says it all.[32]

8. The Centrality of Quality and Its Implications

The present deliberations also serve to indicate, however, why the question of the rate of scientific progress is delicate and rather tricky. For this whole issue turns rather delicately on fundamentally *evaluative* considerations. Thus consider once more the development of science in the recent historic past. As we saw above, at the crudest—but also most basic—level, where progress is measured simply by the growth of the scientific literature, there has for centuries been the astonishingly swift and sure progress of an exponential growth with a doubling time of roughly fifteen years. At the more sophisticated and demanding level of high quality results of a suitably "important" character, there has also been exponential growth—although only at the pace of a far longer doubling time, perhaps thirty years, approximating the reduplication with each successive generation envisaged by Henry Adams at the turn of this century.[33] Finally, at the maximally exacting level of the really crucial insights—significant scientific knowledge that fundamentally enhances our understanding of nature—our analysis has it that science has been maintaining a merely constant pace of progress.

It is this last consideration that is crucial for present purposes, carrying us back to the point made at the very outset. For what we have been dealing with is that essentially seismological standard of importance based on the question, "If a certain finding were abrogated or abandoned, how extensive would the ramifications and implications of this circumstance be? How great would be the shocks and tremors felt throughout the whole range of what we (presumably) know?" What is at issue is exactly a kind of cognitive Richter Scale based on the idea of successive orders of magnitude of impact. The crucial determinative factor for increasing importance is the extent of seismic disturbance of the cognitive terrain. Would we have to abandon or rewrite the entire textbook, or a whole chapter, or a section, or a paragraph, or a sentence, or a mere footnote?

So, while the question of the rate of scientific progress does indeed involve the somewhat delicate issue of the evaluative standard that is at issue, our stance here can be rough and ready—and justifiably so, because the precise details do not affect the fundamental shape of the overall evaluation. *Viewing science as a cognitive discipline*—a body of knowledge whose task is the unfolding of a rational account of the modus operandi of nature—we have it that progress stands correlative with its accession of really major discoveries: the seismically significant, cartography-revising insights into nature. This has historically been growing at a rate that is sure and steady—but essentially linear. However, this perspective lays the foundation for our present analysis—that the historical situation has been one of the constant progress of science as a *cognitive discipline,* notwithstanding its exponential growth as a *productive enterprise* (as measured in terms of resources, money, manpower, publications, etc.).

In the struggle for cognitive control over nature, we have been confronting an enterprise of ever escalating demands. So, while one cannot hope to predict the *content* of future science, the F/I relationship does actually put us into a position to make plausible estimates about its *volume*. To be sure, there is, on this basis, no inherent limit to the possibility of future progress in scientific knowledge. The exploitation of this theoretical prospect, however, gets ever more difficult, expensive, and demanding in terms of effort and ingenuity. New findings of equal significance require ever greater aggregate efforts. In the ongoing course of scientific progress, the earlier investigations in the various departments of inquiry are able to skim the cream, so to speak: they take the easy pickings, and later achievements of comparable significance require ever deeper forays into complexity and call for ever-increasing bodies of information. (It is important to realize that this cost increase is not because latter-day workers are doing *better* science, but simply because it is harder to achieve *the same level* of science: one must dig deeper or search wider to achieve results of the same significance as before.)

A mixed picture emerges. On the one hand, the course of scientific progress is a history of the successive destabilization of theories. On the other hand, the increasing resource requirement for digging into ever deeper layers of complexity is such that successive triumphs in our cognitive struggles with nature are only to be gained at an increasingly greater price. The world's inherent complexity renders the task of its cognitive penetration increasingly demanding and difficult. The process at issue with the growth of scientific knowledge in our complex world is one of drastically diminishing returns. This situation means that grappling with ever more vast bodies of information in the construction of an ever more cumbersome and complex account of the natural world is the unavoidable requisite of scientific progress.

To be sure, we constantly seek to "simplify" science, striving for an ever smaller basis of ever more powerful explanatory principles, but in the course of this endeavor we invariably complicate the structure of science itself. We secure greater power (and, as it were, *functional* simplicity) at the price of greater complexity in *structural* regards.[34] By the time the physicists get that grand unified theory, the physics they will have on their hands will be complex to the point almost of defying comprehension. The mathematics gets ever more elaborate and powerful, the training transit to the frontier ever longer. So, despite its quest for greater operational simplicity (economy of principles), science itself is becoming ever more complex (in its substantive content, its reasonings, its machinery, etc.). Simplicity of process is here more than offset by complexity of product. The phenomenon of diminishing returns reflects the price that this ongoing complexification exacts.[35]

5 Against Convergentism

Synopsis

(1) Some theorists maintain a diminishing-returns view of scientific discovery and envisage a convergentism that holds that later issues are lesser issues. (2) This view faces grave difficulties, for scientific change is not a matter of making minor adjustments in earlier views. Natural science does not develop by accretion and cumulation but by way of substitution and replacement. (3) The equilibrium of science is always unstable. Science not only grows but changes; new science not only supplements but also abrogates the old. The conceptual innovation correlative with these revisions also assures an unending prospect for scientific progress. (4) The potential limitlessness of natural science—the prospect of its ever-continuing progress—does not require an ontological assumption of the infinite structural complexity of nature seeing that endless functional or operational complexity would certainly suffice. (5) Even the unending functional complexity of nature is not required, because the crux is not nature itself but how we inquirers manage to deal with it. Responsibility for the open-endedness of science need not lie on the side of nature at all but can rest one-sidedly with us, its explorers. (6) Natural science is not a body of knowledge secured once and for all, but represents an ongoing process of innovation that engenders an ever-changing picture of the nature of things. A uniform level of significance is generally maintained across these changes.

1. The Diminishing-Returns View of Scientific Progress and Its Flaws

Even if one accepts Kant's Principle of Question Propagation to the effect that in resolving our present scientific questions new ones always arise, the prospect nevertheless remains that the *magnitude of the issues* might grow smaller and smaller as science progresses. Later questions, it might be held, are always smaller questions, so that later science is always lesser science. Successive innovation becomes a matter of increasing refinement in detail and furnishes new materials whose inherent significance decreases continually—exactly as with the decimal elaboration of pi.

Scientific inquiry would thus be conceived of as analogous to terrestrial exploration, whose product—geography—yields results of continually smaller significance that fill in ever-more minute gaps in our information. In such a view, the later investigations yield findings of ever-smaller importance, with each successive accretion making a relatively smaller contribution to what has already come to hand. The advance of science leads, step by diminished step, toward a fixed and final view of things.

This general position is central to Charles Sanders Peirce's vision of ultimate convergence in scientific inquiry:

> As the investigation goes on, additions to our knowledge ... are of less and less worth. Thus, when Chemistry sprang into being, Dr. Wollaston, with a few test tubes and phials on a tea-tray, was able to make new discoveries of the greatest moment. In our day, a thousand chemists, with the most elaborate appliances, are not able to reach results which are comparable in interest with those early ones. All the sciences exhibit the same phenomenon.[1]

In this accretional view of the progress of science, each successive addition is taken to represent a relatively smaller contribution in comparison to what has already been achieved. Successive questions are of lesser and lesser magnitude, continually decreasing in significance and descending to ever more fine-grained levels of detail.

Peirce's view of scientific progress underwrites the provocative step equating "*the real* truth" with "*our ultimate* truth"—the result that science will approximate asymptotically in the long run.[2] Invoking the idea of "ultimate science" or "science in the limit"

$$S_\infty = \lim_{t \to \infty} S_t$$

Peirce then proceeded to construe truth-in-S_∞ as definitive.

Such a position of course has its appeal. Aristotle maintained that the human mind stands in horror before the prospect of a nonconvergent infinite regress (*regressus in infinitum*) in the cognitive domain. Yet it stands in no less horror before the prospect of unending infinite progress (*progressus in infinitum*). It is all too tempting to think that indefinitely prolonged inquiry must in the end achieve a final and definite result.

The Peircean approach has the great theoretical advantage of portraying the real truth not as a relation between our "domesticated" knowledge and something that lies wholly and altogether outside this sphere but rather as something determinable *within* the sphere of our knowledge. Truth is no longer a matter of relating the manifold of our knowledge S_t to an external reality but merely a matter of having the successive S_t stages themselves converge toward a limit. Absolute truth is brought down to earth.

Peirce's espousal of this ultimate-approximation theory of asymptotic development toward a position of fixity made him the most eloquent and certainly the deepest exponent of the diminishing-returns model of scientific knowledge. His strategy deserves closer examination.

2. A Critique of the Self-Correction Thesis

Peirce's theory of ultimate convergence is bound up with his belief in the ultimately self-corrective nature of the scientific method. This reflects a view that long antedates Peirce himself. Over the years, various theorists have argued that natural science is self-corrective—that it develops in such a way that its later stages will, in due course, inevitably correct whatever errors may have arisen earlier on. This doctrine goes back to several eighteenth-century scientists (especially David Hartley, 1705–1757; Georges Le Sage, 1724–1803; and Joseph Priestley, 1735–1804), who took as their descriptive model the various mathematical methods of successive approximation, exemplified by such procedures as the well-known process for the arithmetical calculation of n-th roots.[3] These methods—the rule of false position, for example—begin with guesswork (however wild), providing an *automatic* procedure for successively refining and developing a wrong answer into one that moves stage by stage more closely toward the correct answer.[4] As Priestley put it:

> Hypotheses, while they are considered merely as such, lead persons to try a variety of experiments, in order to ascertain them. These new facts serve to correct the hypothesis which gave occasion to them. The theory, thus corrected, serves to discover more new facts, which, as before, bring the theory still nearer to the truth. In this progressive state, or method of approximation, things continue.[5]

Le Sage drew the analogy between the work of the scientist and that of an arithmetician working a problem in long division, producing in the quotient at each stage a number more accurate than that of the preceding stage.[6] Implicit throughout this eighteenth-century view is the conception of science as possessing an *automatic, mechanically routine method* for improving in the future on its older performance.

This conception of science as automatically self-corrective cannot be sustained. For one thing, when an accepted hypothesis becomes untenable, neither the scientific method nor any other cognitive resource at man's disposal affords any automatic device for producing a new hypothesis that can appropriately replace it. Moreover, the idea of the eighteenth-century theorists (reflected also in Peirce) that science proceeds by way of successive approximation to the truth—that scientific progress leads successively ever closer to the truth by way of an asymptotic approach to "the final answer"—encounters grave obstacles. Apart from its internal difficulties relating to the concept of successive approximation,[7] it lies wholly beyond any prospect of confirmation. Given that we have no science-independent means for determining the truth about nature, we clearly have no prospect of substantiating the claim that our scientific claims are drawing nearer to "the real truth."

It may seem tempting to say that later theories simply provide localized readjustments and that the old theories continue to hold good provided only that we suitably restrict their domains of purported validity. On such a view, it is tempting to say: "Einstein's theory does not *replace* Newton's; it does not actually disagree with Newton's at all but simply sets limits to the region of phenomena (large-scale, slow-moving objects) where Newton's theory works perfectly well." Such temptations must be resisted. To yield to them is like saying that "All swans are white" is true all right; we just have to be cautious about its domain limitation and take care not to apply it in Australia. This sort of position comes down, in the final analysis, to the unhelpful truism that a theory works where a theory works. In science we do not seek local theories that unaccountably hold for limited parametric ranges, but global theories that accountably take special forms within delimited ranges. And just this makes scientific theories vulnerable.

Newton's editor Roger Cotes maintained (in his preface to the second edition of the *Principia*) that the Newtonian system need "assume no principle not proved by phenomena," and that while others had contemplated gravitation before Newton, he "was the only and the first that could demonstrate it from appearances and make it a solid foundation."[8] In this perspective, standard in the seventeenth century, natural philosophy becomes a kind of cryptanalysis. God impresses certain laws of operation upon nature, and we exploit the texts at our disposal (the "phenomena") to decode them. Once we find the key, it is effectively self-evident that we've got it right. In "natural philosophy," as in decipherment (Bentley's recovery of the digamma of Homeric Greek, for example), we manage to achieve a result that, once obtained, is secured forever. This is a pretty picture, but unfortunately it falsifies the way in which things actually work in scientific inquiry.

Even two centuries after Newton, Robert A. Millikan, the American Nobel prize winner in physics, wrote that "in science, truth once discovered always remains truth."⁹ Until a generation ago, students of scientific method were deeply committed to the view that science is cumulative and indeed tended to regard the progressiveness of science in terms of this cumulativity.[10] But in recent decades, this view has come under increasingly sharp attack—and rightly so. The history of science is one long litany of abandoned "truths," including:

- The crystalline spheres of ancient and medieval astronomy
- The humoral theory of medicine
- The effluvial theory of static electricity
- "Catastrophist" geology, with its commitment to a universal (Noachian) deluge
- The phlogiston theory of chemistry
- The caloric theory of heat
- The vibratory theory of heat
- The vital force theories of physiology
- The electromagnetic ether
- The luminiferous ether
- The theory of circular inertia
- Theories of spontaneous generation.[11]

Nor are abandonments of this sort a thing of the primitive past:

> We all know, of course, about "caloric" and "phlogiston" and "spontaneous generation"; but those, we say were childish fallacies of a past era, and not characteristic of modern science, where the errors are local and short-lived. Consider, however, a scientific fallacy on the largest possible scale, which persisted for half a century, until about 1960, in the face of widely publicized counter-arguments and counter-evidence. It is no longer necessary, surely, to remind the reader that Alfred Wegener put forward the hypothesis of *continental drift* in 1912 on the basis of the excellent fit of the continental margins. If this fit were not an extraordinary coincidence, then further evidence supporting the hypothesis was worth consideration. For example, the distribution of various animal and plant species in the southern hemisphere could be explained by radiative dispersion across Africa and South America as if the Atlantic Ocean did not exist. Wegener pointed, also, to geological evidence such as similar geological formations in North-East Brazil and West Africa that might once have been joined. In many respects, Wegener's book is very convincing. Why, then, was his hypothesis rejected by the great majority of geologists for something like 50 years? How did the enormous fallacy of the fixity of the main continental land masses remain entrenched in geology for so long—accompanied, incidentally, by such fantasies

as "land bridges" that eventually sank without trace into the wide, deep water of the South Atlantic? The story is complicated.... For our present purpose it is enough to draw attention to this striking example of a large scale fallacy persisting for half a century in a mature and sophisticated field of science.[12]

Scientific theories at the frontier of research have a notoriously short half-life. Our current theories do not replace old error by secured truth but by *conjectures* regarding the truth that will—so we fully expect—eventually come to be seen as gravely defective. In science there is not, and cannot be, any warrant for holding that our present view of the matter is definitive and will, in essentials, stand secure for all time to come.

The defect of the Peircean theory of asymptotic convergence emerges in this light. For one thing, the successive appropriation theory sees scientific progress as a whole in terms of one particular (and by no means typical) sort of progress, namely, the sequential filling-in of certain basically fixed positions in greater and greater detail—the working out of more decimal places to lend additional refinement to a roughly predetermined result. Neither the historical facts nor the theoretical general principles of the situation lend support to such a theory that scientific progress is a matter of diminishing differences in the meaning-content of the theories at issue. There seems precious little reason to think that the actual progress of science is such that the successive stages of knowledge will come to be less and less distant from one another. As Thomas Kuhn, Paul Feyerabend, and others have argued, there is no reason to think that the succession of theories in science actually involves a historical convergence in the theoretical accounts given.

Another key problem relates to the existence of such a limit. The very idea of a limiting process hinges on the formidable difficulty of how one can possibly define a metric to measure the "distance" between bodies of knowledge.[13]

Moreover, even if there was a limit, we could not find out. For we ourselves inevitably occupy the short run, while limits have to do with what happens in the eventual long run, and, as Keynes said, "in the long run we shall all be dead." Anything we can actually get hold of is without any implications, one way or the other, for what will happen "in the limit."

Any such picture of convergence, however carefully crafted, will also shatter against the *conceptual innovation* that continually brings entirely new, radically different scientific concepts to the fore and brings in its wake an ongoing wholesale revision of established fact. Consider how many facts about a simple object—a sword, for example—were unknown to the ancients. They did not know that it contained carbon or that it conducted electricity. The very concepts at issue ("carbon," "electricity conduction") were outside their cognitive range. There are key facts (or presumptive facts) even about the most familiar things—trees and animals, bricks and mortar—that were unknown a hundred years ago. This ignorance arises because the required concepts have not been formulated. It is not just that the scientists of antiquity did not know what the half-life of *californium* is but that they couldn't have understood this fact if someone had told them about it. The fact is that in natural science small innovations in the data can constrain large changes in their theoretical systematization.

Peircean convergentism accordingly involves massive difficulties. Ongoing scientific progress is emphatically not simply a matter of increasing accuracy by extending the numbers in our otherwise stable descriptions of nature out to a few more decimal places. It will not serve to take the preservationist stance that the old theories are generally acceptable as far as they go and merely need supplementation; significant scientific progress is genuinely revolutionary in that there is *a fundamental change of mind* as to how things happen in the world. Progress of this caliber is generally a matter not of adding further facts—on the order of filling in a crossword puzzle—but of changing the framework itself. The fact is that in natural science small innovations with respect to data can constrain large changes in theoretical systematization.

3. The Instability of Science: The Role of Conceptual Innovation

Prominent contemporary philosophers of science of otherwise the most divergent orientations—including Feyerabend, Kuhn, Lakatos, and Quine—agree with rare unanimity that the traditional theory of cumulation simply does not work.[14] Karl Popper is the father of this line of thought: "It is not the accumulation of observations which I have in mind when I speak of the growth of scientific knowledge, but the repeated overthrow of scientific theories, and their replacement by better or more satisfactory ones."[15] And this is surely right. The medicine of Pasteur and Lister does not add to that of Galen or of Paracelsus; it replaces rather than supplements or improves it.

Science progresses not just additively but also subtractively. Today's most significant discoveries generally represent a revolutionary overthrow of yesterday's: the current big findings of science invariably contradict its earlier big findings.

The later positions of science give no allegiance to the concepts or theories of their predecessors—indeed, not even to their data. For the very data of natural science can always be abandoned. The shift from "Why is p the case?" (Why does the sun go around the earth?) to "Why did it seem to people that p is the case?" always threatens. Even our "observations" can come to be reclassified as mere appearances—as "phenomena" in the old sense. Theory may reshape the very evidence of our senses.

Progress in basic natural science is a matter of constantly rebuilding from the very foundations. Significant scientific progress is generally a matter not of adding further facts but of changing the framework itself; substantial headway is made preeminently by conceptual and theoretical innovation.

The problems solved by one theory may be left unsolved (and even unraisable) by its successors. If we abandon Cartesian vortices, Newton's critics asked, how can we explain the unidirectional relation of the sun's planets? It is a telling obstacle to the doctrine of scientific cumulativity that the whole history of science is one long story of tidy solutions becoming unraveled in the course of progress. One recent discussion cites a plethora of illustrations:

> That phenomenon is illustrated within physics by the failure of Newton's optics to solve the problem of refraction in Iceland spar (which had been explained by Huygens' optics), and by the failure of early nineteenth-century caloric theories of heat to explain phenomena of heat convection and generation; problems which had been solved by Count Rumford in the 1790s. Within chemistry, many problems which had been solved by the early theories of elective affinity were not solved by Dalton's later atomistic chemistry. A still better example is afforded by Franklinic Electrical theory. Prior to Franklin, one of the central solved problems for electricity was the mutual repulsion of negatively charged electrical bodies. Various theories, especially vorticular ones, had solved this problem by the 1740s. Franklin's own theory, which was widely accepted from the middle to the end of the eighteenth century, never adequately came to grips with this problem.[16]

Relativity brushes aside questions regarding the fine structure of the electromagnetic ether (as investigated by Kelvin, Larmor, and Boltzmann). Its rivals unceremoniously discarded the concept of chemical "affinities," whose degrees could now no longer be investigated. Science in the main develops not by way of cumulation but by way of substitution and replacement.[17]

The scientific theories of one era bear little resemblance to those of another. The equilibrium achieved by natural science at any given stage of its development is always an unstable one. Scientific theories have a finite life span. They come to be modified or replaced under various innovative pressures, in particular the enhancement of observational evidence (through improved experimental technique, more powerful means of observation, superior procedures for data processing, etc.).

A state of the art of natural science is a human artifact; and like all other human creations, it has a finite life span. As a creation within time, the ravages of time will wear it away. What the course of scientific development brings into being, it will also take away. Accordingly, *our* science simply is not in a position to deliver a definite picture of physical reality. There is no reason to think that, in the future, scientific theorizing must in principle reach a final and definitive result. Scientific work at the creative frontier of theoretical innovation is always done against the background of the realization that anybody's "findings"—one's own included!—will eventually be abandoned and be superseded by something rather different. What is stable about science is its basic aims, not its questions—let alone its answers!

The adequacy of a scientific thesis is not just a matter of what somebody or many people or everyone at a certain time *thinks* is correct—but what really and truly *is* so. Yet the history of the subject shows that natural science can certify nothing. Definitive and certifiedly correct answers to scientific questions are something that neither past nor present science offers and that future science will not offer either. Only perfected science can provide this—and perfected science is something we do not and doubtless never shall have.

Why does X occur in circumstance Y? Because Y activates the boojums, and boojums always produce X. Science can always answer any explanatory or descriptive question. Anybody can. But that, of course, is not the point. What matters is not just having *an* answer but having the *correct* answer. And even this is not quite right. Asked "What's the core temperature of the earth?" you reply, "$(10^7 \pm 8,500)$°C," pulling numbers out of thin air. It's your lucky day, and it turns out by chance that you are right (so let us suppose). Splendid! But still not good enough. For what matters is not just having an answer that is correct but having one that can be established as tenable and appropriate in the circumstances.

It will not serve to take the preservationist stance that the old views were acceptable as far as they went and merely need supplementation and refinement. Significant scientific progress is genuinely revolutionary and involves *fundamental change of mind* as to how things happen in the world. Relativity theory does not amend the doctrine of luminiferous ether but abandons it. The creative scientist is every bit as much a demolition expert as a master builder.

While natural science may aspire to deliver the certifiable, definitive truth about the world and its ways, this goal is in principle unattainable at the level of generality and precision at which science operates. Our scientific findings are always subject to the realization that our successors a thousand years hence may (quite justifiedly) think very differently than we do ourselves. It is altogether rational to accept the well-established theories of natural science; but not because we know them to be true. We do no such thing, realizing full well that they will one day be superseded. Rather, we accept them *faute de mieux*, because they represent the best available answers to our questions—the most warrantable conjectures we can make here and now. The relation between "scientific warrant" and "actual truth" involves a gap that we can never satisfactorily bridge. We can speak unproblematically of "the best estimate of the truth that we can make in the present state of the art," but we can never lay confident claim to the truth as such in matters of scientific thinking.

Surely, we are told, science makes progress by finding out more truth. Why else is later science better science? "The alternative to accepting that there is a strong measure of truth in science is to go back to blaming a witch when the cow is sick."[18] But this view is profoundly wrong. It is correct that later science is better established science. It is correct that it would be irrational to exchange later science for earlier science (let alone superstition!). But the superiority of later science does not lie in its having more truth or being closer to the truth. Rather, it lies in its affording us a more defensible supposition regarding the truth—in its affording us theses whose acceptance is better warranted despite the fact that we abstain from claims about their actual relationship to the real truth (more of it, closer to it, or the like). We cannot relate scientific appropriateness to truth *directly* without an epistemic detour into the warrant (evidence, justification) for a *claim* to truthfulness.

The idea of science as able to reach final and definitive results is perhaps the last vestige of the classical conception of science of the seventeenth and eighteenth centuries. The idea that reason, proceeding from a vast body of empirical observations, will in the end discern a single definitive structure reflects the grand aspiration of classical metaphysics: that reason in nature will deliver unto reason in the minds of inquirers a clear, determinate, all-embracing formula. The collapse of this metaphysical doctrine is part and parcel of a tilt of the balance between reason and experience that has proceeded throughout the development of modern science. If we see interactive experience as the key to all knowledge of nature, then we are bound to see whatever is based (as science ever must be) on an invariably limited experience-in-hand as something tentative and imperfect. Ernan McMullin has put the point trenchantly: "So many are the veils of language and hypothesis that lie between us and the real, that we now tend to believe that something exhaustible must be something man-made, or, correlatively, that the real can never be conceptually exhausted."[19]

In science, substantial headway is made preeminently by conceptual and theoretical innovation. And this innovation makes our entire earlier picture of natural phenomena look naive, a matter of crude attempt and imperfect understanding—or grave misunderstanding. The fact that this innovation is conceptual also means that certain crucial issues have previously remained wholly outside the cognitive range of earlier inquirers. Projecting this into the future, we are led to the conclusion that there will always be explanatorily significant facts (or plausible candidate-facts) about a thing that we ourselves do not know because we cannot even *conceive* of them. For to grasp such a fact means taking a perspective of consideration that we simply do not have, since the state of knowledge (or purported knowledge) is not yet advanced to a point at which its formulation is possible.

Why should we not take the stance that the movement of scientific knowledge is substantively directional—that science may indeed not have reached the ultimate truth but is drawing nearer to it? The reply is that this idea of "drawing closer to a definitive picture of reality" rests on a fallacy. It proposes to move from the fact that we have better warrant for accepting a revised picture of nature to the conclusion that this picture is closer to the truth (more faithful)—that is, it moves from a picture-of-nature's being a *better warranted picture* to its being a *better picture*. This is a slide we cannot make, for the later theories of science are superior to the earlier not because of their more faithful delineation of nature but because they afford us enhanced means of prediction and control.

Given that science, in the main, develops not by way of cumulation but by way of substitution and replacement, it is unreasonable to think of the development of science in terms of convergence or approximation. In matters scientific, not only are we not in a position to claim that we are achieving the truth, we cannot even establish that we are drawing *closer* to it. For how could we possibly do so without knowing where it lies? We have no reason to think that we are not exchanging one incorrect opinion for another. It is an important fact of epistemology—perhaps the single most important fact of the field—that we have no satisfactory way of securing definitive information about the world at the level of scientific generality and precision.

The successive stages of science do indeed come to be more and more successful in their realization of the aims of the enterprise; in some sense, they do an increasingly more adequate job. But this increasing success is attained across a never-lessening range of content diversity. Our commitment to the view that later theories are better theories does not justify the idea of an ever closer approach to a position of fixity. Improvements in grounding do not engender improvements in depiction: a better evidentiated contention does not thereby become truer, if only because probability of truth is something very different from distance to the truth.

Such a position sees validating the claims of science in terms of the truthfulness of its products as picking up the wrong end of the stick: one should not approach warrant by way of truthfulness, but truthfulness by way of warrant. The claims of later science are to be taken to afford better estimates of the truth because of their enhanced epistemic grounding: they are ruled to be more acceptable by methods of inquiry whose credentials—relative to all envisaged alternatives—are substantially superior.

Scientific progress thus correlates with warranted confidence. But the augmentation in our level of confidence moves—at the level of theory *content*—through a series of radical changes and discontinuities. (The situation is *not* one of the Peircean picture of stable essentials, of improvements only at the decimal points.) With the transition from a corpus of scientific knowledge to a successor, we get neither more of the truth nor draw nearer to it; rather, we get a better-based estimate of the truth—one that gives us a firmer warrant for our claims. Accordingly, while we can make method-oriented claims about "improving correctness," we should make no such claims at the contentual, substance-oriented level of "increasingly close approximation to a 'correct picture of things.'" Our talk of progress cannot bypass the epistemic and regulative aspect and move directly to a descriptive and constitutive result. To reemphasize: one must resist any temptation to think of improvement in *warrant* in terms of improvement in *approximation*.

4. The Potential Limitlessness of Scientific Change

Some theorists see science as an essentially closed venture that will ultimately come to the end of its tether. One recent author suggests that all that is requisite for science to answer every question in principle answerable by the scientific method is that the number of questions be finite and that science answer more and more questions in the course of time.[20]

To articulate this idea, let us contemplate the prospect of resorting to the definition:

M = the set of all (legitimate) scientific questions—all that are "empirically meaningful," in something like the logical positivist sense of being answerable in principle by the scientific method.

Big problems arise in contemplating this set **M**. The very idea of **M** is problematic in its wholly unrelativized reference to "all scientifically meaningful questions." There simply is no absolute, state-of-the-art, independent way of denominating a particular question as scientifically meaningful. (This, at bottom, is the reason why it makes no sense to think of completeness of science in terms of an ability to resolve "all scientific questions" or "all questions in principle answerable by the scientific method.")

Once we reintroduce the needed state-relativization, the situation is drastically altered. For then the claim that finitude entails exhaustibility is refuted by the consideration that the number of (significant) scientific questions may be finite within every S_t and that, nevertheless, even if more and more questions are answered with increasing t' there will never be a time when all scientific questions are resolved. For even if the size of $Q(S_t)$ is always finite (for any value of t), but yet grows with t' then the size of the body of actually answered questions $Q^*(S_{t'})$ can grow *ad indefinitum* without producing any diminution in the pool of questions still confronting us unresolved at any given stage. Indeed, if we can answer 60 percent of an ever-growing (but ever finite) body of questions, the relative size of our "sphere of ignorance" will never diminish, while its real size will represent an ever-growing quantity.

Still, is it the case that no matter how far science advances, there is yet further important work to be done—that we shall never run out of new frontiers for exploration?

The eminent biologist Bentley Glass made newspaper headlines in December 1970 with his presidential address to the American Association for the Advancement of Science, posing the question "Are there finite limits to scientific understanding, or are there endless horizons?"[21] His answer ran as follows:

> What remains to be learned may indeed dwarf imagination. Nevertheless, the universe itself is closed and finite.... The uniformity of nature and the general applicability of natural laws set limits to knowledge. If there are just 100, or 105, or 110 ways in which atoms may form, then when one has identified the full range of properties of these, singly and in combination, chemical knowledge will be complete. There is a finite number of species of plants and of animals—even of insects—upon the earth. We are as yet far from knowing all about the genetics, structure and physiology, or behavior of even a single one of them. Nevertheless, a total knowledge of all life forms is only about $2 \times 10g$ times the potential knowledge about any one of them. Moreover, the universality of the genetic code, the common character of proteins of different species, the generality of cellular structure and cellular reproduction, the basic similarity of energy metabolism in all species and of photosynthesis in green plants and bacteria, and the universal evolution of living forms through mutation and natural selection all lead inescapably to a conclusion that, although diversity may be great, the laws of life, based on similarities, are finite in number and comprehensible to us in the main even now. We are like the explorers of a great continent who have penetrated of its margins in most points of the compass and have mapped the major mountain chains and rivers. There are still innumerable details to fill in, but the endless horizons no longer exist.[22]

This perspective sees the scientific project as a potentially bounded venture, subject to the idea that in scientific inquiry, as in geographical exploration, we are ultimately bound to arrive at the end of the road.[23]

This position is seemingly a minority view. For even with the currently fashionable stress on man's limitations, many theorists do not think that nature has a finite cognitive depth. They view nature as cognitively inexhaustible: we can (in theory) always learn more and more about it, attaining ever-new horizons of discovery, with the new no less interesting or significant than the old. But how can unlimited scientific discovery be possible?

To underwrite the prospect of endless progress in the discovery of natural laws, some theorists have felt compelled to stipulate an intrinsic infinitude in the structural makeup of nature itself.[24] The physicist David Bohm, for example, tells us that "at least as a working hypothesis science assumes the infinity of nature; and this assumption fits the facts much better than any other point of view that we know."[25] Bohm and his congeners thus postulate an infinite quantitative scope or an infinite qualitative diversity in nature, assuming either a principle of unending intricacy in its makeup or one of unending orders of spatiostructural nesting.

Such a postulate can take various forms. One of them involves the suggestion that the range of natural laws is inexhaustible because of an unending ramification of physical levels in the physico-spatial structure of the world. Scientific progress is seen as a matter of progressively unlocking one world within another. Sometimes this theory of limitless intricacy inclines towards Pascal's idea that nature might be an endless nest of Chinese boxes, with microscopic worlds within ever more minute microcosms.[26] Sometimes this shrinkage toward the submicroscopically small is replaced by an expanding sequence moving unendingly upward toward the infinitely large. Following this line of thought, one might project an ascending sequence: subatomic particles, atoms, molecules, organic microorganisms, animals, populations, life systems, solar systems, galaxies, galactic clusters, worlds, world systems, and so on, each with its own characteristic mode of operation—arranged in a steadily ascending sequence of larger orders of scale, emplacing macroscopic worlds within ever more macroscopic ones. (Perhaps what we see as a galaxy is simply a subatomic particle of a macroworld, and our whole universe no more than an atom.) In either view, the theoretically unending progress of science is assured by the existence of an endless series of new worlds embracing (or embraced by) the old.[27]

Some theorists take a less radical course and exchange an appeal to physical levels of inclusiveness for an ever-deepening succession of operating principles or "forces" within the physical makeup of nature:

> Why should Nature run on just a finite number of different types of force? May there not be an infinite hierarchy of types of force just as there is an infinity of structures that may be built of matter interacting under the influence of any one or more of those forces? . . . So, perhaps, our discovery of a sufficient proliferation of types of force may guide us into an empirical classification of those forces in terms of their relationship to one another—analogous to the grouping of the chemical elements in the Mendeleev table—and then to an understanding of that dissatisfaction in terms of some underlying principle that would then enable us to predict the nature of undiscovered and undiscoverable forces. . . . It is perfectly possible that there exist objects that interact powerfully with each other but only exceedingly feebly with the objects with which we are familiar, that is to say with objects that interact strongly with ourselves, so that these unknown

objects could build up complex structures that could share our natural world but of which we should be ignorant.[28]

Theorists of this persuasion envisage an endless succession of natural forces, each operating at its own parametric level within a cosmic hierarchy. Scientific discovery need never end, because its subject matter, the domain of identifiable physical quantities, is ever widening.

Other theorists, such as C. S. Peirce, see the opening up of new realms of phenomena in evolutionary terms. As the cosmos grows older, novel modes of natural structure gradually develop to afford new phenomena governed by evolving laws of their own. Since the universe affords a panorama of changing modes of physical process, a science that reflects this will always find new grist for its mill.

Still other theorists assume the element of chance and randomness at work in the world and then contemplate a potentially unending hierarchical order of structures governed by principles that represent statistical summaries, as it were, of lower-level tendencies stabilized under the aegis of the law of large numbers. In just this vein, the French physicist Jean Paul Vigier declared:

> We would prefer to say that at all levels of Nature you have a mixture of causal and statistical laws (which come from deeper or external processes). As you progress from one level to another you get new qualitative laws. Causal laws at one level can result from averages of statistical behavior at a deeper level, which in turn can be explained by deeper causal behavior, and so on *ad infinitum*. If you then admit that nature is infinitely complex and that in consequence no final state of knowledge can be reached, you see that at any stage of scientific knowledge causal and probability laws are necessary to describe the behavior of any phenomenon, and that any phenomenon is a combination of causal and random properties inextricably woven with one another.[29]

But is that sort of thing needed? Does the prospect of potentially limitless scientific progress actually require structural infinitude in the physical composition of nature along some such lines? The answer is surely negative.

The prime task of science lies in discovering the laws of nature, and it is law complexity that is crucial for present purposes.[30] Even a mechanism of finitely ramified structure can have endlessly complex laws of operation. For if one remains at a fixed level of scale in physical magnitude, one can still have rearrangements of items, and rearrangements of rearrangements, and rearrangements of rearrangements of rearrangements, *ad infinitum*, with emergent lawful characteristics arising at every stage. The workings of a structurally finite and indeed simple system can yet exhibit an infinite intricacy in *operational* or *functional complexity*, exhibiting this limitless complexity in its workings rather than at the spatio-structural or compositional level. While the number of constituents of nature may be small, the ways in which they can be combined are infinite. Think here of the examples of letters/syllables/words/sentences/paragraphs/books/book families (novels, reference books, etc.)/libraries/library systems. In theory, there is no need to assume a ceiling to such a sequence of levels of integrative complexity. The emergence of new concept concatenations and new laws can be expected at every stage. The different levels each exhibit an order of their own. The laws we attain at the nth level can have features whose investigation lifts us to the $(n + 1)$st. New phenomena and new laws can arise at every new level of integrative order. The different facets of nature can generate new strata of laws that yield a potentially unending sequence of levels, each giving rise to its own characteristic principles of organization, themselves quite unpredictable from the standpoint of the other levels.[31]

A "working hypothesis" of the *structural infinitude* of nature is not needed to assure nonterminating progress in science. An unending depth in the *operational or functional complexity* of nature would be quite enough to underwrite the potential limitlessness of science.

5. The Role of Cognitive Limits

In fact, however, neither unending structural nor operational complexity is required to provide for cognitive inexhaustibility. The usual recourse to an infinity-of-nature principle is strictly one-sided, placing the burden of responsibility for the endlessness of science solely on the shoulders of nature itself. In its view, the potential endlessness of scientific progress requires limitlessness on the side of its *objects*, so that the infinitude of nature must be postulated either at the structural or at least at the functional levels. But this is a mistake.

Science, the cognitive exploration of the ways of the world, is a matter of the *interaction* of the mind with nature—of the *mind's cognitive exploitation of the data to which it gains access* in order to penetrate the secrets of nature. The crucial fact is that scientific progress hinges not just on the structure of nature itself but also on the structure of the information-acquiring processes by which we investigate it.

Suppose a natural system to be such that a certain parameter p cannot be evaluated at the time-point t but only on average during an interval. In such a case, the system can be very simple indeed—it need contain *no* complexities apart from those required to assure the preceding assumptions—yet endless cognitive progress is nevertheless available. For as *our* capacity to make p-determinations down to smaller and smaller time intervals increases from minutes to seconds to milliseconds to microseconds, and so on, we will obtain an increasingly comprehensive insight into the modus operandi of the system and can obtain even fuller information about it. There is no reason to think that the discoveries being made as one moves further down the line are of lesser significance than those achieved early on. The discoveries made at each iteration are alike *major* innovations. If its modus operandi is sufficiently complicated, even a simple system can afford the prospect of an ongoing discovery of facts whose overall significance does not decrease with the succession of stages.

The example of written text in fact provides a helpful analogy for the sort of complexity in cognitive stratification that can underwrite the prospect of unending scientific discovery. Knowing the frequencies of the letters *a* and *t* in a certain group of texts yields virtually nothing as to the frequency of the word *at*. The laws of a given level of discovery need not anticipate or determine those of another. As long as we proceed to analyze written text in more and more sophisticated detail, we can, casting our net more and more widely, constantly uncover new laws. Throughout we remain at one basic level of phenomena—that of text—dealing with one single object from various *perspectives of consideration*. Such an approach to the infinity of nature construes this infinitude not so much in terms of spatial extensiveness or physical structure as of *conceptual depth*.

Note that the complexity of endless hierarchy levels need not enter in through the structural makeup of nature itself but merely in the conceptual apparatus being deployed to study it at ever greater depths of sophistication. The case is similar to that of the geometer, whose hierarchy of definitions—axiomatic facts, lemmas, theorems, and so on—does not reside in his materials as such but in the conceptual taxonomy that he chooses, for cognitive reasons, to impose on these materials. Nature herself has no "depth." For depth (like difficulty) is an inherently relative matter, generated through the operation of a cognitive perspective.[32] The endless levels at issue will not be *physical* levels but *levels of consideration* inherent in the activities of inquiring beings. Complexity, after all, lies less in the objects than in the eyes of their beholder. As Herschel ruminated long ago, particles moving in mutual gravitational interaction are, as we human investigators see it, forever solving differential equations that, if written out in full, might circle the earth.[33]

Ongoing cognitive innovation thus need not be provided for by assuming (as working hypothesis or otherwise) that the system being investigated is infinitely complex in its physical or functional makeup. It suffices to hypothesize an endlessly ongoing prospect of securing fuller information about it. The salient point is that it is *cognitive* rather than structural or operational complexity that is the key here.

If we are sufficiently myopic, then, even when a scene is itself only finitely complex, an ever ampler view of it will come to realization as the resolving power of our conceptual and observational instruments is increased. Thus responsibility for the open-endedness of science need not lie on the side of nature at all but can rest one-sidedly with us, its explorers.

Accordingly, the question of the progressiveness of science should not be confined to a consideration of nature alone, since the character of our information-gathering procedures, as channeled through our theoretical perspectives, is also bound to play a crucial part.[34] Innovation on the side of data can generate new theories, and new theories can transform the very meaning of the old data. This dialectical process of successive feedback has no inherent limits and suffices of itself to underwrite a prospect of ongoing innovation. Even a finite nature can, like a typewriter with a limited keyboard, yield an endlessly varied text. It can produce a steady stream of *new* data—new not necessarily in kind but in their functional interrelationships and thus in their theoretical implications—on the basis of which our knowledge of nature's operative laws is continually enhanced and deepened.

Every time we contrive a different mode of description for actual processes—discover a new branch of mathematics, for example—we find (lo and behold!) that new laws of nature become accessible to our intellect. Who is prepared to say that the realm of pure mathematics—a sphere, presumably, of pure intellectual creativity and inventiveness—represents a finite domain? W. Stanley Jevons wrote in 1873:

> I am inclined to find fault with mathematical writers because they often exult in what they can accomplish, and omit to point out that what they do is but an infinitely small part of what might be done. They exhibit a general inclination, with few exceptions, not to do so much as to mention the existence of problems of an impracticable character. This may be excusable as far as the immediate practical result of their researches is in question, but the custom has the effect of misleading the general public into the fallacious notion that mathematics is a *perfect* science, which accomplishes what it undertakes in a complete manner. On the contrary. . . . Just as the numbers we can count are nothing compared with the numbers which might exist, so the accomplishments of a Laplace or a Lagrange are, as it were, the little corner of the multiplication-table, which has really an infinite extent.[35]

If mathematics is incomplete, so is a natural science that relies in its exploitation for the description of nature. If new mathematics makes new laws of nature accessible to our cognitive grasp, and if the prospect of new mathematics is literally endless, then so, clearly, is the prospect of new science.

The process of cognitive development is such that any law is, potentially, a member of a wider family of laws that will itself exhibit some lawful characteristics and thus be subject to synthesis under "higher" laws. No matter what law may be at issue, we can ask questions about it that demand an answer in lawful terms. Any system is potentially a member of some yet more comprehensive system. This consideration alone blocks any prospect of completing science.

These various considerations combine to indicate that *an assumption of the quantitative infinity of the physical extent of the natural universe or of the qualitative infinity of its structural complexity is simply not required to provide for the prospects of ongoing scientific progress.* Ongoing discovery is as much a matter of how the mind goes to work as it is of the object of inquiry. And this fact constrains us to recognize that even a finitely complex nature can provide the domain for a virtually endless course of new and significant discovery. There is no good reason to think that the natural science of a finite world is an inherently closed and terminable venture, and no adequate basis for the view that the search for greater "depth" in our understanding must eventually terminate at a logical rock bottom.[36]

In science, a better look at things need not add mere points of detail to what has gone before. Given that the transformations from one level of understanding to another can be genuinely revolutionary, there is no reason to think that our later readjustments in theoretical understanding are merely minor addenda to or qualifications of the earlier ones. E. P. Wigner has presented the issue in the following terms:

> The question now comes up whether science will at least be able to continue the type of shifting growth indefinitely in which the new discipline is deeper than the older one and embraces it at least virtually. The answer is, in my opinion, no, because the shifts in the above sense always involve digging one layer deeper into the "secrets of nature," and involve a longer series of concepts based on the previous ones which are thereby recognized as "mere approximation." Thus ... first ordinary mechanics had to be replaced by quantum mechanics, thus recognizing the approximate nature and limitation of ordinary mechanics to macroscopic phenomena. Then, ordinary mechanics had to be recognized to be inadequate from another point of view and replaced by field theories. Finally, the approximate nature and limitation to small velocities of all of the above concepts had to be uncovered. Thus, relativistic quantum theory is at least four layers deep; it operates with three successive types of concepts, all of which are recognized to be inadequate and are replaced by a more profound one in the fourth step. This is, of course, the charm and beauty of the relativistic quantum theory and of all fundamental research in physics. However, it also shows the limits of this type of development. The recognizing of an inadequacy in the concepts of the tenth layer and the replacing of it with the more refined concepts of the eleventh layer will be much less of an event than the discovery of the theory of relativity was.[37]

Why should the successive "layers" of physical theory be seen as increasingly less important, less interesting, or less valuable? The discovery of the need to reconstrue the concepts of the i-th layer in terms of those of the $(i + 1)$st is surely every bit as significant for large as for small values of i. No preestablished harmony—no metaphysical hidden hand—is at work to assure that the *order of discovery* in our penetration of the secrets of nature replicates their *order of importance*, so that later penetrations of the "deeper" layers are always of lower orders of cognitive significance for the constituting of an adequate picture of nature. Quite to the contrary, any overall restructuring of our view of nature should be seen—on balance—as having an importance equal to any other. The fact is that progress in natural science carries us through a series of transformations and revolutions whose *overall* significance is essentially uniform. However, the substantiation of this important point raises large issues that demand separate consideration.

6. Scientific Changes Maintain a Uniform Level of Significance

Science is actually not so much a system, a finished structure of knowledge, as a process—an inquiring activity whose ultimate goal may indeed be the completion of a finished and perfected system but which proceeds in the full recognition that this aim is ultimately unreachable. We have to accept the idea that while *progressing* makes sense, *arriving* does not. We cannot attain perfection; we can always do more, but we cannot do it all. As John Herschel wrote in 1831: "[Science is] essentially incomplete, and incapable of being fully embodied in any system or embraced by any single mind ... [so that] in whatever state of knowledge we may conceive man to be placed, his progress towards a yet higher state need never fear a check but must continue till the last existence of society."[38]

Neither theoretical issues of general principle nor the actualities of historical experience suggest that scientific progress need ever come to a stop.[39] Of course, it is possible that for reasons of exhaustion, of penury, of discouragement, we humans might cease to push the frontiers forward. But should we ever abandon the journey, it will be for reasons such as these and not because we have reached the end of the road.

Since scientific progress is a matter of replacement rather than mere supplementation, there is no reason to see the *later* issues of science as *lesser* issues in the significance of their bearing upon science as a cognitive enterprise—to think that nature will be cooperative in always yielding its most important secrets early on and reserving nothing but the relatively insignificant for later on. (Nor, on the other hand, does it seem plausible to think of nature as perverse, leading us ever more deeply into deception as inquiry proceeds.) A very small-scale effect, even one that lies very far out along the extremes of a range exploration in terms of temperature, pressure, velocity, or the like—can force a far-reaching revision and have profound impact by way of major theoretical revisions. (Think of special relativity in relation to ether-drift experimentation, or general relativity in relation to the perihelion of Mercury.)

Thus, we envisage neither a convergent nor a divergent series here, but pretty much of a steady-state condition in which every major successive stage in the evolution of science yields innovations, and innovations of roughly equal overall interest and importance.[40]

The issue that comes to the fore here is reminiscent of the late nineteenth-century controversy between W. Stanley Jevons and those who, like John Tyndall and Charles Sanders Peirce, saw the history of science as addressing itself to issues of ever decreasing magnitude. Thus Jevons wrote:

> Professor Tyndall . . . likens the supply of novel phenomena to a convergent series, the earlier and larger terms of which have been successfully disposed of so that comparatively minor groups of phenomena alone remain for future investigators to occupy themselves upon. On the contrary, as it appears to me, the supply of new and unexplained facts is divergent in extent, so that the more we have explained the more there is to explain. The further we advance in [developing] any generalisation, the more numerous and intricate [and significant] are the exceptional cases still demanding further treatment. . . . By . . . [numerous] illustrations it might readily be shown that in whatever direction we extend our investigations and successfully harmonise a few facts, the result is only to raise up a host of other unexplained facts.[41]

The most plausible view of the importance of scientific innovation surely lies in the middle ground between a theory of diminishing returns and a theory of ever-widening new worlds to conquer, holding that the successive stages of science pose issues of relatively stable *overall* importance for our understanding of the world. Given that natural science progresses not so much by way of improvements as by substitutions and replacements that go back to first principles and lead to a comprehensive overall revision of our picture of the phenomenon at issue, it seems sensible to say that the shifts across successive scientific "revolutions" maintain the same level of overall significance. At the cognitive level, a scientific innovation is simply a matter of change; it is neither a convergent nor a divergent process.[42] In aggregate, the course of change across successive scientific revolutions does not move us from the more important to the less so, or the reverse, but maintains a uniform level of overall significance.

6 Question Dynamics and Problems of Scientific Completeness

Synopsis
(1) There can be no such thing as an all-purpose predictor. (2) Among other reasons for this are the essentially intractable problems of self-prediction.

1. The Impracticability of an All-Purpose Predictive Engine

The predictive domain provides some of the clearest available indications of our limitations in matters of question resolution. The conditions of human life being what they are, many of our most urgent and intriguing questions relate to the future. Nevertheless, we here encounter difficulties that are deeprooted in the nature of our epistemic situation. It is instructive to consider the grounds and implications of our incapacity in the face of predictive questions.

Special-purpose instrumentalities for prediction regarding industrial production, earthquakes, the weather, demography, and the like are a familiar part of the present-day scene. But let us imagine the project of combining all such resources together in one great, all-encompassing interlinkage, implementing the idea of one unrestrictedly versatile and synoptically effective predictive resource. We thus envision creating an All-Purpose Predictive Engine capable—in principle—of resolving any predictive question that one might want to consider.

Let us now suppose that such a project is pursued and issues in the production of the (hypothetical) all-purpose prediction machine *Pythia*. And let us further assume that with the passage of time this machine is ongoingly enhanced and improved. Indeed, let us contemplate the prospect that it is *perfected*. Just what would this mean? How are we to understand perfection in this context of an all-purpose predictor?

The "perfection" at issue here will have two principal aspects: performing without errors of commission (without the incompetence-indicative failure of making incorrect predictions), and performing without errors of omission (without the impotence-indicative failure of an inability to resolve meaningful predictive questions). We may characterize these two factors as reliability perfection and versatility perfection, respectively. Let us explore their ramifications.

Suppose that we have an early model of Pythia at hand and that its record of predictive success is respectable but by no means flawless. We ponder: "Will a totally reliable Pythia be developed by the year 2,500—one that is error-free, so that it never makes an incorrect prediction?" We can certainly wonder about this predictive question. It is clear that a negative answer here could eventually be disconfirmed when that future Pythia malfunctions. *But a positive answer can never be validated.* We can, of course, evidentiate this claim to perfection to some extent though a record of predictive success, but we can never actually verify it in the sense of credibly establishing its truth. For—obviously—the prospect of failure with a yet-unasked question can never be set aside on the basis of information actually obtainable by us at any given time. The prospect of Pythia's luck—or, rather, competence—running out at some point can never be securely excluded.

There is, to be sure, the intriguing prospect of putting to Pythia itself the predictive question: "Are you reliability-imperfect—that is, will you ever at some point in the future come up with an incorrect prediction?" It is clear that we could gain but little added confidence from a reassuringly negative answer here. Future error would now just transmute into present error as well. Of course we have no way to check up on it. Claims to reliability perfection are always indecisive. They can never be satisfactorily substantiated irrespective of whether they are made by us (the manufacturers and users of Pythia) or by Pythia itself.

Now on to versatility perfection. Could Pythia predict *everything*? Could it provide a tenable answer for every predictive question one might care to ask? Clearly not. For one thing, there are essentially paradoxical questions on the lines of: "What is a predictive question that you will never, ever have to deal with in answering questions put to you?" Pythia itself must keep silent on this issue—or else confess its versatility imperfection with an honest "can't say." But the question itself is certainly not meaningless, since we ourselves or some other predictor could in principle answer it correctly.

Again, we could wire our supposedly perfect predictor up to a bomb in such a way that it would be blown to smithereens when next it answers "yes." And we now ask it: "Will you continue to be active and functioning after giving your next answer?" Should it answer "yes," it would be blasted to bits and its answer thereby falsified. Should it answer "no" it would *(ex hypothesi)* continue in operation and its answer thereby be falsified.[1] Viability can be achieved here only by letting the predictor have the option of responding "can't say" at its disposal—and thereby fail to be versatility-perfected. For some predictive questions are unresolvable as a matter of practical reality. ("Where will the last man ever to think about Napoleon be born?") Others are unresolvable as a matter of general theory. (For example, "How long will you need to resolve this number-theoretic question?" asked of a computer in the context of the arithmetical questions at issue with Alan Turing's intractable "halting problem," is thus unresolvable.)[2] No predictor is immune from failings.

To say that a predictive question is unresolvable is not, of course, to say that a predictor cannot answer it. We could certainly program a predictive machine to answer every yes/no question by "yes," every *name* question by "George Washington," every *when?* question by "in ten minutes," and so on. If it is simply an answer that we want, we can contrive predictors that will oblige us. But the problem, of course, is to have a predictor whose predictions can lay a cogent claim to being credible in advance of the fact.

There certainly are predictive questions that will baffle the very best of predictors. Specifically, one of the issues regarding which no predictor can ever function perfectly is its own predictive performance. For consider asking that putatively perfect predictor the question: "Will you answer this question—this *very* question—indefinitely, that is, by 'can't say' rather than yes or no?" We arrive here at the following situation:

The answer given is	Its truth status is automatically	So the predictor
YES	false	fails
NO	false	fails
CAN'T SAY	false	fails

Our supposedly perfect predictor will inevitably fail to function adequately since in no case can it possibly provide a correct answer to such questions.

The long and short of it is that no predictor can succeed with respect to its own operations and, in particular, none can predict its own future predictions.[3] The very most one could plausibly ask for might seem to be that *predictors predict correctly everything that is in principle possible for them to predict.* But even here there are problems. Perhaps the most obvious of them is that of deciding just what is "in principle possible" for a predictor by way of genuine prediction. To determine what is in theory or in principle predictable—and what is not—we have to look to the deliverances of natural science. Science itself tells us that some sorts of predictions are in principle infeasible—for example, the outcome of such quantum processes as are at issue in the question: "Exactly when will this atom of a transuranic element with half-life of 283 years disintegrate?" "Exactly what will be the position and momentum of this particle two minutes from now?" What is needed at this point is a determination of *what is in fact predictable*—and not just an indication of what the science of the day considers (perhaps mistakenly) to be predictable. This is an issue that could only be settled satisfactorily by perfected or completed science, something we do not have now and could not be sure of having even if (per impossible) we did sometime acquire it.[4] We have to come to terms with the frustrating but unavoidable circularity inherent in the holistic aspect of cognitive justification: predictive perfection would require scientific or explanatory perfection, and such perfection could only be validated via predictive perfection.

The predictive inaccessibility of the future of the world (as best we can comprehend its nature) is clearly something that is not *merely* epistemic, something that does not *just* inhere in our own lack of information—our own ignorance. Rather, to the best of our understanding it roots in the very nature of things in a world whose dynamical development is subject to contingency. It is an aspect of reality as such—a result of the fact that presently existing conditions always encompass genuine contingency in that nature simply "has not yet made up its mind" about the future—"has not yet decided" at the present exactly and completely what that future is going to be like. The prominent role on the world's stage of prediction-precluding factors such as chance, choice, and chaos prevents the future from being somehow comprehended in the present. It is this factual feature of the world (as best we can now comprehend it) that blocks the path to achieving perfection in matters of prediction.

Yet another key predictive limitation pertains not to the suppliers of predictions but their recipients. For even if (contrary to hypothesis) we had actually perfected an error-free predictive engine and moreover (again contrary to hypothesis) could know this to be so, we nevertheless would be unable to make adequate use of it. A question-resolving resource is no better than a questioner who uses it. Thus suppose that there had been at Julius Caesar's disposal a predictive machine that could answer questions about the outcome of U.S. congressional elections in the twentieth century—one that gave totally reliable information about the outcome—candidate by candidate—in every district. Clearly this resource would have done Caesar no good at all because it would not—and could not—occur to him to ask the questions at issue. The very concepts involved in the description of certain future developments—biographical, social, scientific, mathematical, or whatever—are unavailable to those who live and think prior to the innovations themselves. Our conceptions limit our cognitive horizons and conceal various regions of future occurrence in an unavoidable darkness of unknowing. We can only ask where we can imagine, and our imagination is circumscribed by the cognitive resources and limitations of our time and setting.

2. Problems of Reflexivity and Metaprediction

Seeing that it makes no sense to ask a predictive engine to predict everything, it seems sensible to inquire whether such an instrumentality could predict everything that is in fact predictable—everything that could reasonably and appropriately be asked of some predictor or other. Alas, even here there are problems.

As we have already seen, metapredictions are predictions about predictions. These can meaningfully arise in many different ways, as for example with: "I predict that no American president will ever be able to predict the identity of his/her 20th successor," or, "I predict that meteorologists will eventually be able to forecast changes of climate one hundred years in advance." But special problems of legitimacy can arise with *reflexive* metapredictions—predictions about classes of predictions to which the prediction at issue itself belongs. For such predictive reflexivity can lead to unusual, literally extraordinary situations. For example, it can engender self-refuting predictions, as per: "No one will ever ask (or wonder) whether Julius Caesar ever had the hiccups." In making this prediction I automatically falsify it. On the other hand, predictions can also be reflexively self-verifying. Thus consider "I predict that by the end of the day someone will have made a silly prediction"—a prediction that provides for its own truth.

Let us return for a moment to our hypothetical predictive machine Pythia. Difficulties are clearly going to arise as long as Pythia's performance is itself part of the domain regarding which it is being asked to make predictions. For then there will, on grounds of theoretical general principle, be some issues with respect to which Pythia cannot function appropriately. Consider an example. The questions that one asks Pythia can be divided into two groups, the *anticipated* (which Pythia has at some prior point predicted it would be asked), and the *unanticipated* (which Pythia has never predicted before the occasion itself that it would be asked). Now suppose we now ask Pythia "What will be the next unanticipated predictive question that you will be asked?" It is clear that Pythia will be unable, in principle, to give a correct answer. While *we* ourselves—or some *other* predictor—might perfectly well predict correctly here, Pythia itself certainly cannot.

Moreover, it is readily demonstrated that every predictor (natural or artificial) has blind spots: it must sometimes answer "can't say" even where *another* predictor can quite appropriately respond yes or no. This circumstance emerges from the following predictive question:

(Q) Will the register of predictive questions that you, Pythia, will resolve affirmatively this year ultimately *omit* (that is, fail to include) this very question Q itself?

Clearly, this question will drive Pythia into perplexity. Suppose the question is answered "yes." Then Q is a predictive question that Pythia answers affirmatively and that must therefore itself feature on the register at issue—which is exactly what that affirmative answer denies. So in this case, Pythia has answered our predictive question incorrectly. On the other hand, suppose that the question is answered "no." Then, given the substance of Q, the register of affirmatively resolved predictive questions will have to omit Q, so that, in consequence, the proper answer is "yes." Here too, Pythia has answered our predictive question incorrectly. (The situation is summarized in table 6.1; note that the first and last columns systematically conflict here.) With no prospect of a correct answer in the yes/no range, Pythia's best—indeed only—error-avoiding option is to make the incapacity-acknowledging response "can't say." (In this way, every predictor is unavoidably versatility-imperfect.) But observe that this is *not* an issue on which rational prediction is inherently impossible. We ourselves could perfectly well give an affirmative answer—and could well be correct in doing so. It is again essential to the paradoxical aspect of our hypothetical situation that this question of predictive performance is addressed to the predictor itself; reflexivity is a crucial factor here.

Again, consider asking our predictor the question: "Will you answer the *next* question that you are asked affirmatively?" As it stands, this is clearly a perfectly cogent, self-consistent question. But consider this question in the context where:

1. If we get a YES answer, we propose to ask next: "Will the next question you will be asked complete a series of three negative answers?" (This is a question the predictor must answer NO, thereby falsifying its previous answer.)

2. If we get a NO answer, we propose to ask: "Will your answer to this question be the same as the answer to the previous one?" (This is a question the predictor stands committed to answering negatively—thereby incapacitating itself from giving a correct answer.)

Either way, we have forced the predictor into error within the setting of this interactive situation. But we have not done this by posing a question that is improper in itself. Another predictor can in principle resolve it correctly; it is only inappropriate for the particular predictor at issue.

Table 6.1. A Survey of Possible Responses to Q.

Answer given to Q	Is Q on the register according to this answer?	Answer implied by Q's inclusion/omission
YES	NO	not YES
NO	YES	YES

Note: By hypothesis, a question is to be on the register if the answer given to it is YES.

Reflexivity is the crux here. The long and short of it is that every predictor is bound to manifest versatility-incapacities with respect to its own predictive operations. A "perfected predictor" that correctly answers every meaningful and answerable predictive question—those regarding its own operations included—is a logico-conceptual impossibility.

7 The Unpredictability of Future Science

Synopsis

(1) Might there be questions that science is simply impotent to resolve? (2) It is difficult, indeed impossible, to predict the future of science. For we cannot forecast in detail even what the questions of future science will be—let alone the answers. (3) Viewed not in terms of its aims but in terms of its results, science is inescapably plastic: it is not something fixed, frozen, and unchanging but something endlessly variable and protean—given to changing not only its opinions but its very form. (4) We thus cannot discern the substance of future science with sufficient clarity to say just what it can and cannot accomplish. (5) Setting domain limitations to science—putting entire ranges of phenomena outside its explanatory grasp—is a risky and unprofitable business.

1. Difficulties in Predicting Future Science

As the preceding discussion has shown, any predictive resource is bound to encounter serious difficulties when employed reflexively in being asked to make predictions about its own performance. Self-prediction is a profoundly problematic issue.

This is nowhere more decidedly the case than with respect to natural science. The prospect for making scientifically responsible predictions about the future of science itself is deeply problematic. The splendid dictum that "the past is a different country—they do things differently there" has much to be said for it. We cannot fully comprehend the past in terms of the conceptions and presumptions of the present. And this is all the more drastically true of the human future—in its cognitive aspect in particular. After all, information about the thought world of the past is at any rate available—however difficult extracting it from the available data may prove to be. But it lies in the nature of things that we cannot secure any effective access to the thought world of the future. Its science specifically included, all its details are hidden from our view. All that we know is that it will be different.

The idea of chance in innovation is one major source of scientific cognition's unpredictability. Entire fields of empirical inquiry have been launched by serendipity. After the American Telegraph and Telephone Company started its overseas shortwave radio telephone service in 1927, Karl Jankey, a Bell Laboratories scientist, began to monitor shortwave static. He observed one kind of static whose source he could not identify but whose intensity peaked four minutes earlier each sidereal day. Jankey's surprisingly regular "noise" was composed of radio signals from astronomical objects. Thus a quality control project for the telephone industry launched the innovative field of radio astronomy, which opened the way for investigating astronomical features and processes undetectable at optical wavelengths. (Serendipity pervades radio astronomy: Penzias and Wilson, also scientists from Bell Laboratories, were using their radio telescope to track communications satellites when they picked up—and initially failed to recognize—cosmic background radiation. This finding lent important empirical support to the Big Bang theory and so fostered innovation in theoretical cosmology.)[1]

In particular, those who initially conceive and produce a device (a typewriter, say, or a printing press, or even a guillotine) cannot possibly foresee all the uses to which it will eventually be put. Negative prediction may be possible—there will certainly be things one cannot do with it (you cannot use a screwdriver as a fountain pen—although you could write with it in the manner of the Babylonian tablets). But it is often difficult to map out in advance the range of things that *can* be done with a mechanism—and next to impossible to specify in advance the range of things that *will* be done with it. Human ingenuity is virtually limitless, and the old devices are constantly put to new and previously unimaginable uses. One must, after all, never forget the prospect of major innovation issuing from obscurity. The picture of Einstein toiling in the patent office in Bern should never be put altogether out of mind. Commenting shortly after the publication of Frederick Soddy's speculations about atomic bombs in his 1930 book *Science and Life*,[2] Robert A. Millikan, a Nobel laureate in physics, wrote that "the new evidence born of further scientific study is to the effect that it is highly improbable that there is any appreciable amount of available subatomic energy to tap."[3] In science forecasting, the record of even the most qualified practitioners is poor.

The best that we can do in matters of science and technology forecasting is to look toward those developments that are in the pipeline by looking to the reasonable extrapolation of the character, orientation, and direction of the current state of the art. This is a powerful forecasting tool on the positive side of the issue. And this conservative approach has its problems. Since we cannot predict the answers to the presently open questions of natural science, we also cannot predict its future questions. For these questions will hinge upon those as yet unrealizable answers, since the questions of the future are engendered by the answers to those we have on hand. Accordingly, we cannot predict science's solutions to its problems because we cannot even predict in detail just what these problems will be.

Of course, once a body of science comes to be seen as something settled and firmly in hand, many issues become routine. Various problems become mere reference questions—a matter of locating an answer within the body of pre-established, already available information that somewhere contains it. (The *mere* here is, to be sure, misleading in its downplaying of the formidable challenges that can arise in looking for needles in large haystacks.) However, in pioneering science we face a very different situation. People may well wonder about the cause of a certain phenomenon—what causes cancer, say, or what produces the attraction of the lodestone for iron—in circumstances where the concepts needed to develop a workable answer still lie beyond their grasp. Indeed, most of the questions with which present-day science grapples could not even have been raised in the state of the art that prevailed a generation ago.

In scientific inquiry as in other sectors of human affairs, major upheavals can come about in a manner that is sudden, unanticipated, and often unwelcome. Major breakthroughs often result from research projects that have very different ends in view. Louis Pasteur's discovery of the protective efficacy of inoculation with weakened disease strains affords a striking example. While studying chicken cholera, Pasteur accidentally inoculated a group of chickens with a weak culture. The chickens became ill but, instead of dying, recovered. Pasteur later reinoculated these chickens with fresh culture—one strong enough to kill an ordinary chicken. To Pasteur's surprise, the chickens remained healthy. Pasteur then shifted his attention to this interesting phenomenon, and a productive new line of investigation opened up. In empirical inquiry, we generally cannot tell in advance what further questions will be engendered by our endeavors to answer those on hand, those answers themselves being as yet unavailable. Accordingly, the issues of future science simply lie beyond our present horizons.

The past may be a different country, but the future is a terra incognita. Its science, its technology, its fads and fashions lie beyond our ken. We cannot begin to say what ideas will be at work here, although we know on general principles they will differ from our own. And where our ideas cannot penetrate we are ipso facto impotent to make any detailed predictions.

Throughout the domain of inventive production in science, technology, and the arts we find processes of creative innovation whose features defy all prospects of predictability. The key fact in this connection is that of the fundamental epistemological law: *the cognitive resources of an inferior (lower) state of the art cannot afford the means for foreseeing the operations of a superior (higher) one.* Those who know only tic-tac-toe cannot foresee how chess players will resolve their problems.

We know—or at any rate can safely predict—that future science will make major discoveries (both theoretical and observational/phenomenological) in the next century, but we cannot say what they are and how they will be made (since otherwise we could proceed to make them here and now). We could not possibly predict now the substantive content of our future discoveries—those that result from our future cognitive choices. For to do so would be to transform them into present discoveries which, by hypothesis, they just are not.[4] In the context of questions about matters of scientific importance, then, we must be prepared for surprises.

It is a key fact of life that ongoing progress in scientific inquiry is a process of conceptual innovation that always places certain developments outside the cognitive horizons of earlier workers because the very concepts operative in their characterization become available only in the course of scientific discovery itself. (Short of learning our science from the ground up, Aristotle could have made nothing of modern genetics.) What one scientific generation sees as a natural kind, a later one disassembles into a variety of different species. We have as yet no inkling of the concept mechanisms that later scientific eras will make use of. The major discoveries of later stages are ones that the workers of a substantially earlier period (however clever) not only have failed to make but which they could not even have understood, because the requisite concepts were simply not available to them. Newton could not have predicted findings in quantum theory any more than he could have predicted the outcome of American presidential elections. One can only make predictions about what one is cognizant of, takes note of, deems worthy of consideration. Thus, it is effectively impossible to predict not only the answers but even the questions that lie on the agenda of future science. For new questions in science always arise out of the answers we give to old ones (as per Kant's Principle of Question Propagation). And the answers to these questions involve conceptual innovations. We cannot now predict the future states of scientific knowledge in detail because we do not yet have at our disposal the very concepts in which the issues will be posed.

With respect to the major substantive issues of future natural science, we must be prepared for the unexpected. If there was one thing of which the science of the first half of the seventeenth century was unalloyedly confident, it was that natural processes are based on contact-interaction and that there can be no such thing as action at a distance. Newtonian gravitation burst upon this scene like a bombshell. Newton's supporters simply stonewalled. Roger Cotes explicitly denied there was a problem, arguing (in his preface to the second edition of Newton's *Principia*) that nature was *generally* unintelligible, so that the unintelligibility of forces acting without contact was nothing specifically worrisome. However unpalatable Cotes's position may seem as a precept for science (given that making nature's workings understandable is, after all, one of the aims of the enterprise), there is something to be said for it—not, to be sure, as science but as metascience. For we cannot hold the science of tomorrow bound to the standards of intelligibility espoused by the science of today. The cognitive future is inaccessible to even the ablest of present-day workers. After Pasteur had shown that bacteria could come only from preexisting bacteria, Darwin wrote that "it is mere rubbish thinking of the origin of life; one might as well think of the origin of matter."[5] One might indeed!

The inherent unpredictability of future scientific developments—the fact that inferences from one state of science to another are generally precarious—means that *present-day science cannot speak for future science*. The prospect of future scientific revolutions can never be precluded. We cannot say with unblinking confidence what sorts of resources and conceptions the science of the future will or will not use. Given that it is effectively impossible to predict the details of what future science will accomplish, it is no less impossible to predict in detail what future science will *not* accomplish. We can never confidently put this or that range of issues outside the limits of science, because we cannot discern the shape and substance of future discoveries with sufficient clarity to be able to say with any assurance what it can and cannot do. Present-day natural science cannot speak for future science. Viewed not in terms of its *aims* but in terms of its *results*, science is inescapably *plastic*: it is not something fixed, frozen, and unchanging but endlessly variable and protean—given to changing not only its opinions but its very form.

Not only can one never claim with confidence that the science of tomorrow will not resolve the questions that the science of today sees as intractable, one can never be sure that the science of tomorrow will not endorse what the science of today rejects. This is why it is infinitely risky to speak of this or that explanatory resource (action at a distance, stochastic processes, mesmerism, etc.) as inherently unscientific. Even if X lies outside the range of science as we nowadays construe it, it by no means follows that X lies outside science as such. We must recognize the commonplace phenomenon that the science of the day almost always manages to do what the science of an earlier day deemed infeasible to the point of absurdity ("split the atom," "abolish parity," or the like). With natural science, the substance of the future inevitably lies beyond our present grasp.

In cognitive forecasting, it is the errors of omission—our blind spots, as it were—that present the most serious threat. For the fact is that we cannot substantially anticipate the evolution of knowledge. Given past experience we can feel reasonably secure when we say *that* science will resolve various problems in the future. But *how* it will do so is bound to be a mystery.[6] The fundamental dynamics at work in the cognitive process itself mean that there is no secure way to predict the conceptual and explanatory resources of the future. This fact itself means that no way is open to us to achieve complexities in knowledge or comprehensiveness in understanding. We can always do better. Not only does cognitive progress lead us to results we cannot anticipate, but it is a journey toward a destination we cannot reach.

The fact is that science itself sets the limits to predictability—insisting that some phenomena (the stochastic processes encountered in quantum physics, for example) are inherently unpredictable. This is always to some degree problematic. The most that science can reasonably be asked to do is to predict what it itself sees as in principle predictable—to answer every predictive question that it itself countenances as proper. Thus if quantum theory is right, the position and velocity of certain particles cannot be pinpointed conjointly. This renders the question "What will the exact position and velocity of particle X be at time t?" not insoluble but illegitimate. Question-illegitimacy represents a limit that grows out of science itself—a limit on appropriate questions rather than on available solutions.

In any case, the idea that science might one day be in a position to predict everything is simply unworkable. To achieve this, it would be necessary, whenever we predict something, to predict also the effects of making those predictions, and then the ramifications of those predictions, and so on *ad indefinitum*.

After all, we cannot even say what concepts the inquirers of the future will use within the questions they will raise. Maxwell's research was directed toward answering questions that grew out of Faraday's work. Hertz devised his apparatus to answer questions about the implications of Maxwell's questions. Marconi's devices were designed to resolve questions about the application of Hertz's work. Each confronted a research problem inherent in solutions to earlier problems—a problem that did not even arise before these solutions themselves were established. The continual opening up of new issues means that there is little sense to the idea that the predictive mission of science could ever be pushed through to a final and definitive completion. In addressing predictive questions we can doubtless always improve upon our performance but never perfect it. Here too we confront the inevitability of limitations in the sphere of question resolution. Questions about the cognitive future are as challenging as questions can get to be—and as intractable.[7]

The idea that science has limits can take very different forms. One is that it may run out of steam—that its capacity to raise further major questions may become exhausted. It is this sort of thing that has concerned us in the preceding chapters. Another type of limit is at issue when there are certain important questions or ranges of scientific questions that it is simply impotent to resolve. These two sorts of limits are very different. For it is entirely possible that science might be unlimited in the former way—that is, it can always come upon new and interesting questions to answer—and yet be limited in the latter way in finding some matters intractably beyond its ability to resolve. It is this second sort of limit that now concerns us.

The landscape of natural science is ever changing: innovation is the very name of the game. Not only do the theses and themes of science change but so do the very questions. Of course, once a body of science comes to be seen as something settled and firmly in hand, many issues become routine. Various problems become mere reference questions, a matter of locating an answer within the body of preestablished, already available information that somewhere contains it. (The *mere* here is, to be sure, misleading in its downplaying of the formidable challenges that can arise in looking for needles in large haystacks.) However, in pioneering science we face a very different situation. People may well wonder "what is the cause of X?"—what causes cancer, say, or what produces the attraction of the lodestone for iron—in circumstances where the concepts needed to develop a workable answer still lie beyond their grasp. Scientific inquiry is a creative process of theoretical and conceptual innovation; it is not a matter of pinpointing the most attractive alternative within the presently specifiable range but of enhancing and enlarging the range of envisageable alternatives. Such issues pose genuinely open-ended questions of original research: they do not call for the resolution of problems within a preexisting framework but for a rebuilding and enhancement of the framework itself. Most of the questions with which present-day science grapples could not even have been raised in the state of the art that prevailed a generation ago.

The difficulty of a scientific question—as measured by the resources of time, effort, material resources, and so on that we inquiring investigators must expend in resolving it—can generally not be predicted in advance, save in the case of routine issues. With questions of pioneering research, we usually cannot even begin to predict how much effort we must expend on a solution; this information is generally available only after an answer has been found.

One would certainly like to be in a position to have prior insight into and give some advance guidance to the development of scientific progress. Grave difficulties arise, however, where questions about the future, and in particular the cognitive future, are involved.

When the prediction of a *specific predefined occurrence* is at issue, a forecast can go wrong in only one way: by proving to be incorrect. The particular development in question may simply not happen as predicted. Forecasting *a general course of developments* can go wrong in two ways: either through errors of commission (that is, forecasting something quite different from what actually happens, say, a rainy season instead of a drought), or through errors of omission (that is, failing to foresee some significant part of the overall course-predicting an outbreak of epidemic, say, without recognizing that one's own locality will be involved). In the first case, one forecasts the wrong thing; in the second, there is a lack of completeness, a failure of pre-vision, a certain blindness.

In science forecasting, the record of even the most qualified practitioners is poor. People may well not even be able to conceive the explanatory mechanisms of which future science will make routine use. As one sagacious observer noted over a century ago:

> There is no necessity for supposing that the true explanation must be one which, with only our present experience, we *could* imagine. Among the natural agents with which we are acquainted, the vibrations of an elastic fluid may be the only one whose laws bear a close resemblance to those of light; but we cannot tell that there does not exist an unknown cause, other than an elastic ether diffused through space, yet producing effects identical in some respects with those which would result from the undulations of such an ether. To assume that no such cause can exist, appears to me an extreme case of assumption without evidence.[8]

Since we cannot predict the answers to the presently open questions of natural science, we also cannot predict its future questions. These questions will hinge upon those as yet unrealizable answers, since the questions of the future are engendered by the answers to those we have on hand. Accordingly, we cannot predict science's solutions to its problems because we cannot even predict in detail just what these problems will be.

A degradation relationship thus circumscribes the domain of feasible prediction. The reliability of the claims we can responsibly make declines sharply with the exactness to which we aspire (see figure 7.1). If one insists on resolving a predictive question with high informativeness (exactness, detail, preci-

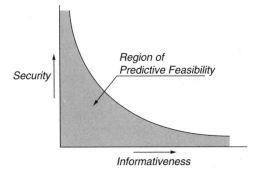

Figure 7.1 The Degradation of Security with Increasing Definiteness

sion), one's confidence in the response should be drastically diminished. Any exercise in science forecasting is bound to conform to this general relationship of futurological indeterminacy.

An ironic but critically important feature of scientific inquiry is that the unforeseeable tends to be of special significance just because of its unpredictability. *The more important the innovation, the less predictable it is, because its very unpredictability is a key component of importance.* Science forecasting is beset by a pervasive normality bias, because the really novel often seems so bizarre. A. N. Whitehead has wisely remarked: "If you have had your attention directed to the novelties in thought in your own lifetime, you will have observed that almost all really new ideas have a certain aspect of foolishness when they are first produced."[9] Before the event, revolutionary scientific innovations will, if imaginable at all, generally be deemed outlandishly wild speculation—mere science fiction, or perhaps just plain craziness.[10]

Thus, while the safe course lies in degrading the informativeness of our forecasts, this obviously undermines their utility—and, in the case of science forecasts, renders them virtually useless. For science scorns the security of inexactness. If we are content to say that the solution to a problem will be of "roughly" this general sort or that it will be "somewhat analogous" to some familiar situation, we will be on safe ground. But science is not like that. It does not deal in analogies and generalities and rough approximations. It strives for exactness and explicitness—for precision, detail, and, above all, for breadth and depth of application. It is just this feature of its claims that makes science forecasting a very risky business. For the more we enrich our forecast with the details of just exactly what and just exactly how (etc.), the more vulnerable it becomes.

In maintaining that future science is inherently unpredictable, one must be careful to keep in view the distinction between substantive and structural issues, between particular individual scientific questions, theses, and theories, and generic features of the entire system of such individual items. At the level of loose generality, various inductions regarding the future of science can no doubt be safely made. We can, for example, confidently predict that future science will have greater taxonomic diversification, greater theoretical unification, greater substantive complexity, further high-level unification and low-level proliferation, increased taxonomic speciation of subject-matter specialties, and so on. We can certainly predict that it will be incomplete, that its agenda of availably open questions will be extensive, and so on. Of course, this sort of information tells us only about the structure of future science, and not about its substance. These structural generalities do not bear on the level of substantive detail: they relate to science as a productive enterprise (or industry) rather than as a substantive discipline (as the source of specific theories about the workings of nature).

With respect to the major substantive issues of future natural science, we must be prepared for the unexpected. We can confidently say of future science that it will do its job of prediction and control better than ours; but we do not—and in the very nature of things cannot—know *how* it will go about this.

2. Present Science Cannot Speak for Future Science

The inherent unpredictability of future scientific developments—the fact that no secure inference can be drawn from one state of science to another—has important implications for the issue of the limits of science. It means that *present-day science cannot speak for future science*: it is in principle impossible to make any secure inferences from the substance of science at one time about its substance at a significantly different time. The prospect of future scientific revolutions can never be precluded. We cannot say with unblinking confidence what sorts of resources and conceptions the science of the future will or will not use. Given that it is effectively impossible to predict the details of what future science will accomplish, it is no less impossible to predict in detail what future science will *not* accomplish. We can never confidently put this or that range of issues outside the limits of science, because we cannot discern the shape and substance of future science with sufficient clarity to be able to say with any assurance what it can and cannot do. Any attempt to set limits to science—any advance specification of what science can and cannot do by way of handling problems and solving questions—is destined to come to grief.

An apparent violation of the rule that present science cannot bind future science is afforded by John von Neumann's attempt to demonstrate that all future theories of subatomic phenomena—and thus all *future* theories—will have to contain an analogue of Heisenberg's uncertainty principle if they are to account for the data explained by present theory. Complete predictability at the subatomic level, he argued, was thus exiled from science. But the "demonstration" proposed by von Neumann in 1932 places a substantial burden on potentially changeable details of presently accepted theory.[11] The fact remains that we cannot preclude fundamental innovation in science: present theory cannot delimit the potential of future discovery. In natural science we cannot erect our structures with a solidity that defies demolition and reconstruction. Even if the existing body of knowledge does confidently and unqualifiedly support a certain position, this circumstance can never be viewed as absolutely final.

The crucial fact is that—basic aims apart—science is not a fixed object but a temporal and transient one; it is invariably a matter of states and stages. We cannot speak atemporally of the facts of science, the questions of science, or the methods of science, flatly and categorically, without a temporal qualification such as "as construed in the seventeenth century" or "as we see them today." The reason for this disability lies exactly in this fact that present-day science cannot speak for future science—that we are not, and never will be, in a position to delineate the issues and materials of *future* science.

How can this be so? Surely "science" cannot change all that drastically without ceasing to be *science*. Well—yes and no. In substantive regards, in its concepts and ideas, natural science is endlessly changeable. Functionally—in terms of the aims of the enterprises encompassing description, explanation, prediction, and control over nature—science indeed is something fixed. The materials of science can vary endlessly, but the range of its definitive tasks is set once and for all.

It is only in functional terms that we can give a satisfying definition of natural science and provide a stable characterization of its nature. The one and only thing that is determinate about science is its mission of description, explanation, prediction, and control of natural phenomena, and the commitment to proceed in these matters by the empirically based rational controls for the testing and substantiation of our assertions that have become known as "the scientific method." Everything else—methods, mechanisms, theories, and so on—is potentially changeable. Matters squarely placed outside the boundaries of science in one era (action at a distance dismissed as unintelligible nonsense, or hypnotism rejected as fakery) can enter within the orthodoxy of a later generation.

An enterprise that, on the cognitive/theoretical side, did not aim at either the description or explanation of nature, and that, on the applied/pragmatic side, did not aim at either prediction or control, simply could not count as natural science. One element on either side may be downgraded, and perhaps even sacrificed. (Instrumentalist physicists have abandoned explanation as a goal of science; the Babylonian astronomer-priests, seeing the celestial bodies as gods, took prediction alone as their pragmatic goal and would have rejected control as blasphemous.) When all the appropriate elements of theoretical and practical purpose are removed, we are not dealing with science at all. This functional basis of ulterior aims is essentially requisite for any enterprise that can be qualified as science. Everything else is potentially changeable. Entire subject matter areas can come (quantum electrodynamics) or go (astrology). Questions can arise (the composition of quarks) or vanish (the structure of the luminiferous ether). Theses and theories can come and go. Methods can evolve. In substance and form alike, science is inherently changeable—indeed, highly volatile.

The epistemological situation of science points toward a theology-reminiscent view of the human condition that sees man as a creature poised between comfortable assurance and the abyss. Our hold on the things of the world—our science included!—is always tenuous. We must view science with the dual realization that it is at once the best we can do and nevertheless by no means good enough. Our scientific picture of the world is vulnerable to being overthrown at any instant by new, unlooked for, and perhaps even unwelcome developments. The history of man's science, like the history of all his works, invites ruminations on the transience of things human and the infeasibility of attaining perfection. (Of course, in our cognitive as in our moral life, the unattainability of perfection affords no reason why we should not do the very best we can; the point is simply that we can never afford the smug confidence that what we have attained is quite good enough.)

At the frontiers of scientific innovation, the halflife of theories and explanatory models is brief. Even particular devices of methodology can change drastically: witness the rise of probabilistic argumentation and use of statistics in the "design of experiments." Larry Laudan has illustrated this point nicely:

> In many cases, *it is the methodology of science itself which is altered*. Consider, as but one example, the development of Newtonian theory in the eighteenth century. By the 1720's, the dominant methodology accepted alike by scientists and philosophers was an inductivist one. Following the claims of Bacon, Locke, and Newton himself, researchers were convinced that the only legitimate theories were those which could be inductively inferred by simple generalization from observable data. Unfortunately, however, the direction of physical theory by the 1740's and 1750's scarcely seemed to square with this explicit inductivist methodology. Within electricity, heat theory, pneumatics, chemistry, and physiology, Newtonian theories were emerging which postulated the existence of imperceptible particles and fluids—entities which could not conceivably be "inductively inferred" from observed data, . . . and so, various Newtonians (e.g., LeSage, Hartley, and Lambert) insisted *the norms themselves should be changed* so as to bring them into line with the best available physical theories. This latter group took it on themselves to hammer out a new methodology for science which would provide a license for theorizing about unseen entities. (In its essentials, the methodology they produced was the hypotheticodeductive methodology, which even now remains the dominant one.) This new methodology, by providing a rationale for "microtheorizing," eliminated what had been a major conceptual stumbling block to the acceptance of a wide range of Newtonian theories in the mid and late eighteenth century.[12]

The quintessential requisites of scientific intelligibility can themselves undergo drastic transformation: witness the fall of the hallowed proscription of action at a distance in the age of Newtonian physics. Science can and does change the rules of the game on us. It can and does abandon whatever is unserviceable and coopt whatever serves its purposes.

Any attempt to specify the "essential presuppositions" or "unchangeable commitments" of science is fruitless. The principle that every event has a specific cause whose operation explains why it eventuated thus rather than otherwise has been an article of scientific faith since classical antiquity. "Nihil turpius physico fieri quidquam sine causa dicere," said Cicero,[13] and for two millennia this doctrine that "nothing occurs by chance" remained a received dogma in science until quantum theory came along to sweep this "indispensable" principle of causality away with a single stroke.

R. G. Collingwood valiantly attempted to specify certain absolute (or ultimate, or primal) presuppositions of the scientific enterprise.[14] He held, in particular, that the uniformity of nature must be presupposed if the scientific enterprise is to succeed. But who is to say that cosmology may not decide tomorrow that the universe is partitioned into distinct compartments (and/or eras) within which different ground rules apply—that is, that the "laws of nature" in the universe are not uniform? What seems an absolute presupposition of science at one point may be explicitly denied at another. In postulating the changeless universality of these "absolute presuppositions," Collingwood's doctrine involves him in an unsubstantiatable (and uncharacteristic) absolutism.

Natural science is altogether ruthless and opportunistic—very much a fairweather friend. If an old and heretofore useful theory can no longer do serviceable work, science has no hesitancy about dropping it. If a heretofore rejected theory proves helpful in altered circumstances, science has no hesitancy about taking it up. On the side of its procedures and theses, science has no fixed nature, no stable commitments; it is prepared to turn whichever way the wind blows. Science is fickle: it is given to flirtations rather than lasting relationships. It is prepared on any and every day to wipe the slate clean if this should prove to be advantageous.

Science is incurably pragmatic and settles for whatever works, abandoning its established general principles with a shameless disloyalty when changed circumstances recommend this course. When something new comes along that proves more effective than existing "knowledge," science immediately changes course. Success is the guiding star: science immediately coopts whatever succeeds in the sphere of its mission. If something other than existing science were to work out better, science itself would change to embrace it. We cannot permanently exclude something from the boundaries of science, because these boundaries are changeable, which is to say that they are effectively nonexistent.

The supremacy of natural science is closely bound up with its plasticity. If we could set limits to the shape and substance of science, then it could also be possible to set limits to what science can and cannot accomplish. If, for example, we could say (with Emil du Bois Reymond[15]) that the explanatory program of science is an atomistic Newtonianism—a mechanistic world in which we could, at the most and best, achieve the "astronomical knowledge" of a "Laplacean Spirit" who knows the total history of the motions of all particles throughout space for all of time—then we could indeed say (as Reymond does) that science cannot account for the source of motion, the nature of matter, the operations of consciousness, and so on. If we could circumscribe the substantive nature of science, we could also circumscribe its potential achievements. It is just this that is impossible. We cannot defensibly project the present lineaments of science into the future.

While we can say with confidence what the state of science *as we now have it* does and does not allow, we cannot say what science as such will or will not allow. The boundary between the tenable and the untenable in science is never easy to discern. Future science can turn in unexpected and implausible directions. The realm of scientific possibility is unchartable. "There are more things in heaven and earth, Horatio. . . ." That some theoretically available scientific position fails to accord with the science of the day is readily established, but that it is inherently unscientific is something that it lies beyond our powers to show. The contention that this or that explanatory resource is inherently unscientific should always be met with instant scorn. For the unscientific can only lie on the side of process and not that of product; on the side of modes, of explanation and not its mechanisms; of arguments rather than phenomena.

There is indeed a line between science and pseudoscience—and also between competent science and poor science. Yet, such boundaries cannot be drawn on *substantive* grounds. The line between science and pseudoscience cannot be defined in terms of content—in terms of what sorts of theses or theories are maintained—but only in terms of method, in terms of how these theories are substantiated. The inherent unpredictability of change in science is the very hallmark of science. It sets real science apart from the closed structures of pseudoscience, whose methodological defect is precisely the "elegance" with which *everything* falls much too neatly into place. And it means that no sort of idea, mechanism, or issue can be placed with reasonable assurance outside the realm of science as such.

Science, as already noted, is simply too opportunistic to be fastidious about its mechanisms. Eighteenth-century psychologists ruled out hypnotism, nineteenth-century biologists excluded geophysical catastrophes, twentieth-century geologists long rejected continental drift. Many contemporary scientists give parapsychology short shrift—yet who can say that its day will never come? The pivotal issue is not what is substantively claimed by an assertion but rather whether this assertion (whatever its content!) has been substantiated by the prevailing canons of scientific method.

There is, to be sure, rough wisdom in scientific caution and conservatism: it is perfectly appropriate to be skeptical about unusual phenomena constructions and to view them with skepticism. Before admitting strange phenomena as appropriate exploratory issues, we certainly want to check their credentials, make sure they are well attested and appropriately characterized. The very fact that they go against our understanding of nature's ways (as best we can tell) renders abnormalities suspect—a proper focus of worriment and distrust. But, of course, to hew this line dogmatically and rigidly, in season and out, is a mistake. The untenable in science does not conveniently wear its untenability on its sleeve. We have to realize that, throughout the history of science, stumbling on anomalies—on strange phenomena, occurrences that just don't fit into the existing framework—has been a strong force of scientific progress.

One does well to distinguish two modes of strangeness: the exotic and the counterindicated. The *exotic* is simply something foreign—something that does not fall into the range of what is known but does not clash with it either. It lies in what would otherwise be an informational vacuum, *outside* our current understanding of the natural order. (Hypnotism and precognition are perhaps cases in point.) The *counterindicated*, however, stands in actual *conflict* with our current understanding of the natural order (action at a distance for seventeenth-century physics; telekinesis today). One would certainly want to apply rigid standards of recognition and admission to phenomena that are counterindicated, but if and when they measure up, we have to take them in our stride. In science we have no alternative but to follow nature where it leads. And new theories can be ruled out even less than new phenomena, for to do so would be to claim that science has reached the end of its tether, something we certainly cannot and doubtless should not even wish to do.

Throughout natural science, we are poised in a delicate balance between reasonable assurance that what we believe is worth holding to and a recognition that we do not yet have the last word—that the course of events may at any time shatter our best laid plans for understanding the world's ways. We can set no a priori restrictions but have to be flexible. Nobody can say what science will and will not be able to do.

3. Against Domain Limitations

The project of stipulating boundaries for natural science—of placing certain classes of nature's phenomena outside its explanatory reach—runs into major difficulties. Domain limitations purport to put entire sectors of fact wholly outside the effective range of scientific explanation, maintaining that an entire range of phenomena in nature defies scientific rationalization. This claim is problematic in the extreme. The scientific study of human affairs affords a prime historical example. Various nineteenth-century German theorists maintained that a natural science of man—one that affords explanation of the full range of human phenomena, including man's thoughts and creative activities—is in principle impossible. One of the central arguments for this position ran as follows: it is a presupposition of scientific inquiry that the *object* of investigation is essentially independent of the process of inquiry, remaining altogether unaffected. This is clearly not the case, however, when our own thoughts, beliefs, and actions are at issue, since they are all affected as we learn more about them. Accordingly, it was argued, there can be no such thing as a science of the standard sort regarding characteristically human phenomena. One cannot understand or explain human thought and action on the usual procedures of natural science but must introduce some special ad hoc mechanisms for coming to explanatory terms with the human sphere. (Thus, Wilhelm Dilthey, for example, maintained the need for recourse to a subjectively internalized *Verstehen* as a characteristic, science-transcending instrumentality that sets the human "sciences" apart from the standard—that is, natural—sciences.) And, so it was argued, seeing that such a process cannot, strictly speaking, qualify as scientific, there can be no such thing as a "science of man."

This position is badly misguided. With science we cannot prejudge results nor predelimit ways and means. One cannot in advance rule in or rule out particular explanatory processes or mechanisms ("observation-independence," etc.). If in dealing with certain phenomena they emerge as observer-indifferent, so be it; but if they prove to be observer-sensitive, we have to take that in our stride. If we indeed need explanatory recourse to some sort of observer-correlative resource ("sympathy" or the like), then that's that. In science we cannot afford to indulge our a priori preconceptions.

To be sure, if *predictability* is seen as the hallmark of the scientific, then there cannot be a science that encompasses all human phenomena (and, in particular, not a science of science). But, of course, there is no reason why, in human affairs any more than in quantum theory, the boundaries of science should be so drawn as to exclude the unpredictable. Even before the rise of stochastic phenomena in quantum physics, one might have asked: must scientifically tractable phenomena be predictable? Can science not tread where predictability is absent?

When we encounter strange "intractable" or "inexplicable" phenomena, it is folly to wring our hands and say that science has come to the end of its tether. For it is exactly here that science must roll up its sleeves and get to work. Long ago, Baden Powell got the main point right:

> When we arrive at any such seeming boundary of present investigation, still this brings us to no *new world* in which a different order of things prevails; it merely points to what will assuredly be a fresh starting point for future research. It is an unwarrantable presumption to assert, that at a mere point of difficulty or ob-

scurity we have reached the boundary of the dominion of physical law, and must suppose all beyond to be arbitrary and inscrutable to our faculties. It is the mere refuge and confession of ignorance and indolence to imagine special interruptions, and to abandon reason for mysticism.[16]

It is poor judgment to jump from a recognition that the science of the day cannot handle something to the conclusion that science as such cannot handle it.

The changeability and plasticity of science are also a source of its power. Its very instability is at once a limitation of science and a part of what frees it from having actual limits. We can never securely place any sector of phenomena outside its explanatory range.

The upshot of these observations is clear. To set domain limitations to natural science is always risky and generally ill advised. The course of wisdom is to refrain from putting issues outside the explanatory range of science. Present science cannot speak for future science; it cannot establish what science as such can and cannot do. It makes no sense to set limits to natural science itself; the title of this book, *The Limits of Science*, points toward a nonentity. Charles Sanders Peirce's dictum holds good: one must not bar the path of inquiry.

8 Against Insolubilia

Synopsis
(1) That unanswered questions always remain at every stage of scientific development does not mean that there are questions that science cannot answer at all—insolubilia. The immortality of questions in natural science does not imply the existence of immortal questions. (2) Both parties went wrong in the famous du Bois-Reymond/Haeckel Controversy: Reymond was wrong in his claim to have identified insolubilia; Haeckel was wrong in his claim that natural science was effectively complete. (3) A consideration of various purported scientific insolubilia indicates that none of them are tenable as such. None of those "ultimate questions" are inherently unresolvable. (4) While science does indeed have limitations and can never solve all of its problems, it does not have limits. There are no scientifically appropriate questions that science cannot resolve in principle—and thus might not resolve in the future. The quest for scientific insolubilia is a delusion; no one can say in advance just what questions natural science can and cannot manage to resolve. Identifiable insolubilia have no place in an adequate theory of scientific inquiry.

1. The Idea of Insolubilia

Consider the weak limitation on the question-resolving capacity of our scientific knowledge that is characterized by the following thesis:

Weak Limitation: The Permanence of Unsolved Questions

There are *always*, at every temporal stage,[1] questions to which no answer is in hand, questions for whose resolution current science is inadequate, although they may well be answered at some later stage.[2]

This thesis has it that there will *always* be questions that the science of the day raises but does not resolve. It envisions a permanence of cognitive limitation, maintaining that our knowledge is never at any stage completed, because certain then-intractable (that is, posable but as yet unanswerable) questions figure on the agenda of every cognitive state of the art.

Given Kant's Principle of Question Propagation, this permanence of unsolved questions is at once assured. For if every state of knowledge generates further new and as yet unanswered questions, then science will, clearly, never reach a juncture at which all its questions are resolved. The prospect of further progress is ever present: completeness and finality are unrealizable in the domain of science. In this light, the present limitation thesis seems altogether plausible.

It should be noted, however, that weak limitation is perfectly compatible with the circumstance that *every* question that arises at a given stage will *eventually* be answered. The eventual resolution of *all* our (present-day) scientific problems would not necessarily mean that science is finite or completable because of the prospect—nay, certainty—that other issues will have arisen by the time the earlier ones are settled.

A doctrine of perpetual incompleteness in science is thus wholly compatible with the view that every question that can be asked at each and every particular state of natural science is going to be answered—or dissolved—at some future state. For the incompletability of science that follows from Kant's Principle is of a weak sort. It does not support the more startling conclusion that certain scientific questions are *in principle* beyond the reach of science: that there are unanswerable questions placed altogether beyond the limits of possible resolution. Having ever-unanswered questions does not mean having ever-unanswerable ones. We cannot simply transpose the quantifiers of the preceding thesis to obtain the very different thesis of:

Stronger Limitation: The Existence of Insolubilia

There are questions unanswered in every state of science—then-posable questions that cannot be answered then and there.[3]

The weak-limitation thesis envisaged the immortality of questions; this present strong-limitation thesis envisages the existence of immortal questions. This stronger thesis posits the existence of insolubilia, maintaining that certain questions go altogether beyond the limits of our explanatory powers and admit of nonresolution within any cognitive corpus that we are able to bring to realization. This is something very different from the claim of weaker limitation—and far less plausible.

One can also move on to the yet stronger principle of:

Hyperlimitation: The Existence of *Identifiable* Insolubilia

Certain questions can be identified as insolubilia: we can here-and-now formulate questions that can never be resolved and we are able to specify concretely some question that is unanswerable in *any* state of science.[4]

This thesis of hyperlimitation makes a very strong claim—and a distinctly implausible one. After all, even a position holding that there indeed are insolubilia certainly need not regard them as being identifiable in the current state of scientific development. Even if there were actual insolubilia questions that science will never resolve, this would not mean that any such questions can be *specified* by us here and now. The very idea that certain now-specifiable questions can be identified as never to be resolved requires claiming that present science can speak for future science, that the science of today can establish what the science of tomorrow cannot do by way of dealing with the issues. And this, as we have seen, is a contention that is altogether untenable in view of the essential unpredictability of future science.

Let us explore some of the ramifications of the issues raised by this family of insolubility theses.

2. The Reymond-Haeckel Controversy

In the 1880s, the German physiologist, philosopher, and historian of science Emil du Bois-Reymond published a widely discussed lecture on *The Seven Riddles of the Universe* (*Die Sieben Welträtsel*).[5] In it, he maintained that some of the most fundamental problems about the workings of the world were insoluble. Reymond was a rigorous mechanist and argued that the limit of our secure knowledge of the world is confined to the range where purely mechanical principles can be applied. Regarding anything else, we not only *do not* have but *cannot* in principle obtain reliable knowledge. Under the banner of the slogan *ignoramus et ignorabimus* ("we *do not* know and *shall never* know"), du Bois-Reymond maintained a skeptically agnostic position with respect to various foundational issues in physics (the nature of matter and force, and the ultimate source of motion) and psychology (the origin of sensation and of consciousness). These basic issues are simply explanatory insolubilia that altogether transcend man's scientific capabilities. Certain fundamental biological problems he regarded as unsolved but perhaps in principle soluble (although very difficult): the origin of life, the adaptiveness of organisms, and the development of language and reason. As regards his seventh riddle—the problem of freedom of the will—he was undecided.

The position of du Bois-Reymond was soon sharply contested by the zoologist Ernest Haeckel, in a book *Die Welträtsel*, published in 1889,[6] which attained a great popularity. Far from being intractable or even insoluble, Haeckel maintained, the riddles of du Bois-Reymond had all virtually been solved. Dismissing the problem of free will as a pseudoproblem—since free will "is a pure dogma [which] rests on mere illusion and in reality does not exist at all"—Haeckel turned with relish to the remaining riddles. Problems of the origin of life, of sensation, and of consciousness Haeckel regarded as solved—or solvable—by appeal to the theory of evolution. Questions of the nature of matter and force he regarded as solved by modern physics except for one residue: the problem (perhaps less scientific than metaphysical) of the ultimate origin of matter and its laws. This "problem of substance" was the only riddle recognized by Haeckel but was downgraded by him as not really a problem for science. In discovering the "fundamental law of the conservation of matter and force," science had done pretty much what it could do with respect to this problem; what remained was metaphysics, with which the scientist has no proper concern. Haeckel summarized his position as follows:

> The number of world-riddles has been continually diminishing in the course of the nineteenth century through the aforesaid progress of a true knowledge of nature. Only one comprehensive riddle of the universe now remains—the problem of substance.... [But now] we have the great, comprehensive "law of substance," the fundamental law of the constancy of matter and force. The fact that substance is everywhere subject to eternal movement and transformation gives it the character also of the universal law of evolution. As this supreme law has been firmly established, and all others are subordinate to it, we arrive at a conviction of the universal unity of nature and the eternal validity of its laws. From the gloomy *problem* of substance we have evolved the clear *law* of substance.[7]

The basic structure of Haeckel's position is clear: science is rapidly nearing a state in which all big problems admit of solution—substantially including those "insolubilia" of du Bois-Reymond. (What remains unresolved is not so much a scientific as a metaphysical problem.) Haeckel concluded that natural science in its *fin de siècle* condition had pretty much accomplished its mission—reaching a state in which all scientifically legitimate problems were substantially resolved.

The dispute exhibits the interesting phenomenon of a controversy in which both sides went wrong.

Du Bois-Reymond was badly wrong in claiming to have identified various substantive insolubilia. The idea that there are any identifiable issues that science cannot ever resolve has little to recommend it, to be sure, although various efforts along these lines have been made, including:

- The attempts of eighteenth-century mechanists to bar action at a distance
- The attempts of early twentieth-century vitalists to put life outside the range of scientific explicability
- The attempts of modern materialists to exclude hypnotism or autosuggestion or parapsychology as spurious on grounds of scientific intractability.

The historical record does not augur well in this regard. The annals of science are replete with achievements that, before the fact, most theoreticians had insisted could not possibly be accomplished. The course of historical experience runs counter to the idea that there are any identifiable questions about the world (in a meaningful sense of these terms) that do in principle lie beyond the reach of science. It is always risky to say *never*, and particularly so with respect to the prospects of knowledge. Never is a long time, and "never say never" is a more sensible motto than its paradoxical appearance might indicate. The key point is well made by Karl Pearson:

> Now I venture to think that there is great danger in this cry, "We *shall* be ignorant." To cry "We are ignorant" is sage and healthy, but the attempt to demonstrate an endless futurity of ignorance appears a modesty which approaches despair. Conscious of the past great achievements and the present restless activity of science, may we not do better to accept as our watchword that sentence of Galilei: "Who is willing to set limits to the human intellect?"—interpreting it by what evolution has taught us of the continual growth of man's intellectual powers.[8]

However, Haeckel was no less seriously wrong in his insistence that natural science was nearing the end of the road—that the time was approaching when it would be able to provide definitive answers to the key questions of the field. The entire history of science shouts support for the conclusion that even where "answers" to our explanatory questions are attained, the prospect of revision—of fundamental changes of mind—is ever present. For sure, Haeckel was gravely mistaken in his claim that natural science had attained a condition of effective completeness.

Yet both these theorists were, in a way, also right. Bois-Reymond correctly saw that the work of science will never be completed; that science can never shut up shop in the final conviction that the job is finished. Haeckel was surely right in denying the existence of identifiable insolubilia.

3. Some Purported Scientific Insolubilia

Consider some examples of proposed scientific insolubilia:

- Why is there anything rather than nothing? Why are there physical things at all? Why does *anything* exist?
- Why is nature an orderly cosmos? Why are there any natural laws (uniformities, regularities) at all? Why are there causal laws to operate as "the cement of the universe"?
- Granted that there are (perhaps even must be) things and laws, why are they as they are rather than otherwise? Why were the "initial conditions" thuswise, and why are the laws as they are? (For example, why are the laws so orderly rather than more chaotic?)

Notice, first of all, the global and universalistic character of these traditional insolubilia. When we try to answer these questions by the usual device of explaining one thing in terms of another, the former immediately expands to swallow up the latter. The usual form of explanation (subsuming boundary conditions under laws) at once falls into question in a way that makes the explanatory process circular. This totalitarian aspect gives the whole issue a cast that is more philosophical than scientific. The questions at issue relate not so much to the discoveries as to the presuppositions of science.

In theory, four lines of response to such "ultimate questions" for explaining existence are possible:

I. They are illegitimate and improper questions

II. They are legitimate

    1. but unanswerable: they represent a mystery

    2. and answerable

        a. via a substantively causal route that centers on a substance (God) that is self-generated *(causa sui)* and is, in turn, the efficient causal source of everything else.

        b. via a nonsubstantival route that centers on some creatively hylarchic principle that has no basis in some preexisting thing or group thereof.

Alternative I represents a choice of last resort—one that sidesteps rather than confronts the question. Alternative II/1 is not particularly appealing; its recourse to mystery leaves us with the worst of both worlds. Alternative II/2a obviously has its problems in an era where people still continue to endorse the medieval principle that *in philosophia non recurre est ad deum* (roughly: "Don't call on God to pull your philosophical chestnuts out of the fire"). So, considering that all of the other alternatives are unpromising and problematic, II/2b deserves consideration. The idea of a creative principle is at least worth entertaining.

Let us consider more closely some of the issues raised by these three "insoluble" problems, focusing on the question "Why is there anything at all?"

Dismissal of the problem as illegitimate is generally based on the idea that the question at issue involves an illicit presupposition. It looks to answers of the form "Z is the (or *an*) explanation for the existence of things." Committed to this response schema, the question has the thesis "There is a ground for the existence of things; existence-in-general is the sort of thing that has an explanation." This presumption, we are told, is perhaps false.

In principle, this presumed falsity could emerge in two ways:

1. on grounds of deep general principle inherent in the conceptual "logic" of the situation; or

2. on grounds of a concrete doctrine of substantive metaphysics or science that precludes the prospect of an answer—even as quantum theory precludes the prospect of an answer to "Why did that atom of *californium* decay at that particular time?"

Let us begin by considering if the question of existence might be invalidated by considerations of the first sort and root in circumstances that lie deep in the conceptual nature of things. Consider the following discussion by C. G. Hempel:

> Why is there anything at all, rather than nothing? . . . But what kind of an answer could be appropriate? What seems to be wanted is an explanatory account which does not assume the existence of something or other. But such an account, I would submit, is a logical impossibility. For generally, the question "Why is it the case that *A*?" is answered by "Because *B* is the case". . . . *[A]n answer to our riddle which made no assumptions about the existence of anything cannot possibly provide adequate grounds.* . . . The riddle has been constructed in a manner that makes an answer logically impossible.[9]

But this plausible problem-rejecting line of argumentation is not without its shortcomings. The most serious of these is that it fails to distinguish appropriately between the existence of things, on the one hand, and the obtaining of facts, on the other,[10] and supplementarily also between specifically substantival facts regarding existing things, and nonsubstantival facts regarding states of affairs that are not dependent on the operation of preexisting things.

We are confronted here with a principle of hypostatization to the effect that the reason for anything must ultimately always inhere in the operations of things. At this point we come to a prejudgment or prejudice as deep-rooted as any in Western philosophy: the idea that things can only originate from things, that nothing can come from nothing (*ex nihilo nihil fit*), in the sense that no thing can emerge from a thingless condition. Now, this somewhat ambiguous principle is perfectly unproblematic when construed as saying that if the existence of something real has a correct explanation at all, then this explanation must pivot on something that is really and truly so. Clearly, we cannot explain one fact without involving other facts to do the explaining. But the principle becomes highly problematic when construed in the manner of the precept that "*things* must come from *things*," that *substances* must inevitably be invoked to explain the existence of *substances*. For we then become committed to the thesis that everything in nature has an efficient cause in some other natural thing that is somehow its causal source, its reason for being.

This stance is implicit in Hempel's argument. It is explicit in much of the philosophical tradition. Hume, for one, insists that there is no feasible way in which an existential conclusion can be obtained from nonexistential premises.[11] The principle is also supported by philosophers of a very different ilk on the other side of the channel, including Leibniz himself, who writes: "The sufficient reason [of contingent existence] . . . must be outside this series of contingent things, and *must reside in a substance which is the cause of this series*."[12] Such a view amounts to a thesis of genetic homogeneity that says (on analogy with the old but now rather obsolete principle that "life must come from life") "things must come from things," or "stuff must come from stuff," or "substance must come from substance."

Is it indeed true that only things can engender things? Must substance inevitably arise from substance? Even to state such a principle is in effect to challenge its credentials, and this challenge is not easily met. Why must the explanation of facts rest in the operation of *things*? To be sure, fact explanations must have inputs (*all* explanations must). Facts must root in facts. But why thing-existential ones? To pose these questions is to recognize that a highly problematic bit of metaphysics is involved here. Dogmas about explanatory homogeneity aside, there is no discernible reason why an existential fact cannot be grounded in nonexistential ones, and why the existence of substantial *things* cannot be explained on the basis of some nonsubstantial circumstance or principle whose operations can constrain existence in something of the way in which equations can constrain nonzero solutions. Once we give up this principle of genetic homogeneity and abandon the idea that existing things must originate in existing things, we remove the key prop of the idea that asking for an explanation of things in general is a logically inappropriate demand. The footing of the rejectionist approach is gravely undermined.

After all, rejectionism is not a particularly appealing course. Any alternative to rejectionism has the significant merit of retaining for rational inquiry and investigation a question that would otherwise be intractable. The question of "the reason why" behind existence is surely important. If there is any possibility of getting an adequate answer—by hook or by crook—it seems reasonable that we would very much like to have it. There is nothing patently meaningless about this riddle of existence. And it does not seem to rest in any obvious way on any particularly problematic presupposition—apart from the epistemically optimistic idea that there are always reasons why things are as they are (the "principle of sufficient reason"). To dismiss the question as improper and illegitimate is fruitless. Try as we will to put the question away, it comes back to haunt us.[13]

Consider the line of reasoning set out in the list below. Since assertions (A) and (B) squarely contradict each other, it is clear that theses (1-4) constitute an inconsistent group of propositions. One of this quartet, at least, must be rejected.

An Inconsistent Quartet

(1) Everything in nature (macroscopic) has a causal explanation. [The Principle of Causality]

(2) Natural-existence-as-a-whole must itself be counted as a natural thing: the universe itself qualifies as a substance—a thing. [The Principle of Aggregative Homogeneity: the universe consists of things (substances) and is itself a thing (substance).]

(A) The universe has a causal explanation. [From (1) and (2).]

(3) Causal explanations of existential facts require existential inputs to afford the requisite causes. [The Principle of Genetic Homogeneity]

(4) No existential inputs are available to explain the existence of natural-existence-as-a-whole, the totality of things within the world (= the universe). For any existent involved by the explanation would constitute part of the explanatory problem, thus vitiating the explanation on grounds of circularity. [The Principle of Causal Comprehension: anything that stands in causally explanatory connection with the universe is thereby, ipso facto, a part of it.]

(B) No (adequate) causal explanation can be given for the universe. [From (3) and (4).]

Let us explore the options for resolving this inconsistency:

(1)-*rejection*. One could abandon the Principle of Causality. This would pave the way for accepting the universe ("natural-science-as-a-whole") as an item whose existence is uncaused. For obvious reasons, this is not a particularly attractive option.

(2)-*rejection*. One could abandon The Principle of Aggregative Homogeneity and maintain that the assimilation of the universe itself to particular things must be abandoned. Everything-as-a-whole is seen as sui generis and thus not as a literal thing that, along with particular things, can be expected to conform to the Principle of Causality. Accordingly, we would exempt the universe itself from membership in the class of things that have cause. The difficulty with this approach lies in the problem of establishing the grounds of the purported impropriety. We unhesitatingly view galaxies as individual things whose origin, endurance, and nature need explanation. Why not, then, the cosmos as a whole?

(3)-*rejection*. One could reject (3), as we have in fact already proposed to do. Yet in dismissing genetic homogeneity, one would (and should) not abandon it altogether but rather subject it to a distinction. One could then say that there are *two different kinds* of causal explanation: those that proceed in terms of the causal agency of *things* (substance-causality, or S-causality for short), and those that proceed in terms of the causal operation of *principles* (P-causality). Unlike S-causality, P-causality would not require that the causal principle at issue be rooted in the operations of "things." In its preparedness to let laws rather than things exert causal efficacy, the hylarchic principle it envisages would not presuppose any specifically substantival embodiment whatever. Such an approach abandons the deep-rooted prejudice that causal agency must always be hypostatized as the operation of a causal agent. This option envisages a mode of "causality" whose operation can dispense with existential inputs. It recognizes that the orthodox terms of ordinary efficient causality are not the only ones available for developing explanations of existence. Thus, while still retaining the Principle of Causality as per (1), this approach substantially alters its import.

(4)-*rejection*. This course commits us to the idea that existential inputs are available to explain the existence of natural-existence-as-a-whole. Standardly, this involves the invocation of a nature-external, literally supernatural being (God) to serve as the once-and-for-all existential ground in explaining the existence of all natural things. On this *theological* alternative, one would then retain (1) intact by means of the principle that God is *causa sui*. We have already remarked on the methodological shortcomings of this approach. A rational division of labor calls for leaving God to theology and refraining from drafting him into service in the project of scientific explanation.

Each of these solutions exacts a price. Each calls on us to abandon a thesis that has substantial prima facie plausibility and appeal. And each requires us to tell a fairly complicated and in some degree unpalatable story to excuse (that is, to explain and justify) the abandonment at issue.

The point to be emphasized, however, is that (3)-rejection—the recourse, in existence explanation, to a principle that does not itself have an existential grounding in a thing of some sort—emerges as comparatively optimal. The price it exacts, though real, is less than that of its competitors. The consequences it engenders are, relatively considered, on balance the least problematic—which is, of course, far from saying that they are not problematic at all. In the last analysis, we take recourse to P-causality—to the creative operation of principles—*faute de mieux,* because this is the contextually optimal alternative; no better one is in sight. Accordingly, the idea of a hylarchic principle that grounds the existence of things not in preexisting things but rather in a functional principle of some sort—a specifically nonsubstantival state of affairs—becomes something one can at least entertain.

But what could such a hylarchic principle be like?

What is perhaps the most promising prospect takes the form of a teleological "principle of value" to the effect that things exist because "that's for the best."[14] Such a teleological approach would hold that *being* roots in value. To be sure, this leaves a residual issue: "But why should what is fitting exist?" Here one must resist any temptation to say, "What is fitting exists because there is something [God, Cosmic Mind, etc.] that brings what is fitting to realization." This simply falls back into the causal trap. We shall have to answer the question simply in its own terms: "Because that's fitting." Fitness is seen as the end of the line.

To the objection that such an explanation strategy is inherently unscientific, one must coolly reply: "Do tell! From what mountain did your theoreticians' Moses descend with the tablets that say just what sorts of explanatory mechanisms are or are not scientific?" The scientists of the seventeenth century thought gravitational action-at-a-distance absurd. The fashion of the present day could turn out to be just as wrong with respect to teleological explanation. We cannot put anything securely beyond the pale, because we cannot securely say where the boundaries of science are to be located. As we have seen, the science of one era is never in a position to speak for its successors.

This perspective has important implications. It constrains us to recognize that these purportedly intractable questions are "insoluble" not as such but merely *within the orthodox causal framework.* If we take resort to higher ground by expanding or supplementing or replacing this framework, such questions may well become answerable. The fact that the question "Why is there anything at all?" is indeed ultimate for the *framework of efficient causality*—which, given its own nature, cannot come to grips with the issue—does not mean that there may not be some other framework (such as the teleological framework of final causality) that can deal with this issue more or less successfully.[15] Framework-internal ultimacy will not render a question insoluble as such. Such issues bear upon the subsector-internal limitations rather than the more fundamental issue of limits of knowledge. They represent neither unanswerable insolubilia nor improper questions.

The problem of why the initial conditions for the universe are as they are affords yet another candidate insolubile. Some theoreticians regard this issue as lying beyond the reach of scientific explicability. W. Stanley Jevons, for example, has written as follows:

> [Darwin] proves in the most beautiful manner that each flower of an orchid is adapted to some insect which frequents and fertilises it, and these adaptations are but a few cases of those immensely numerous ones which have occurred in the lives of plants and animals. But why orchids should have been formed so differently from other plants, why anything, indeed, should be as it is, rather than in some of the other infinitely numerous possible modes of existence, he can never show. The origin of everything that exists is wrapped up in the past history of the universe. At some one or more points in past time there must have been arbitrary determinations which led to the production of things as they are.[16]

In natural science, so we are told, all we do is make relational assertions of the form that "things are thus at t because they are so at t'" and we are confined to operating on the principles of conditionality to the effect that this is so because that is so. Accordingly, to "explain" the initial conditions of natural processes—the ultimate origins of things—in terms of other states of the system, we would have to move to later times, and this would be unscientific. They have to be viewed as ultimate surds, as "arbitrary determinations" lying beyond the reach of explicability.[17]

Again the question must be pressed: just why is a move to ex post facto explanation inherently unscientific? What is to preclude an explanatory rationalization of prior states in terms of the posteriors to which they lead? Who can legislate a priori just where science may and may not legitimately go—just what explanatory principles it may and may not use? It may well transpire that the increasingly fashionable Anthropic Principle to the general effect that "the initial conditions are as they are because intelligent life otherwise could not have evolved in the universe" involves various errors. Being inherently unscientific, however, is not one of them.[18]

The question "Why are there any laws at all?" is seemingly less pressing than its thing-oriented cousin. After all, laws are an expression of order, and any order (even an emptiness or a chaos) is an order of some specifiable kind. (Even "disorder" is an order of sorts.) Of course, the real question is, "Why are the laws as they are?"—that is, relatively simple, discoverable by creatures of our sort, and so on. To rationalize such lawfulness, one must make the transition to the level of the fundamental structure of the system of nature as a whole.

We thus arrive at that other ultimate question of why the overall systemic structure of nature is as it is—so orderly, simple, intelligible. In the final analysis, this question comes down to: why is the whole systemic order of things and laws as it is rather than otherwise? This leads us back toward the causal difficulties noted above. We cannot give a subsumptive (law-based) explanation here: laws cannot yield unproblematic support to a structure of which they themselves are a part. We must again appeal outside the range of natural law and fact to those creative principles that underlie the entire existential order of nature. We arrive once more at the need for principles of a different order—for example, the teleological: they're the way they are because "that's for the best."

To be sure, the question would now arise: How can such a teleological principle itself be explained—why should optimality exert an existential impetus? The principle itself affords the materials needed for a response. Why should nature be optimific? Because that, too, is for the best. The approach is smoothly self-substantiating—perfectly able to provide an explanation on its own terms, as any holistically systemic explanation would in principle have to do.

Thus the route to insolubilia via ultimate questions can always be blocked. Since ultimacy is never absolute but framework-internal, we can defeat it by shifting to a variant explanatory framework. We must thus recognize these various purported insolubilia for what they are: not really insoluble questions whose resolution lies beyond the explanatory reach of science as such but merely questions whose natural response must be of a sort not particularly congenial to the explanatory preconceptions and prejudices of the day.

4. The Infeasibility of Identifying Insolubilia

As we have seen, any state of science delimits the range of legitimately posable questions to those whose presuppositions concur with its contentions. If quantum theory is right, the position and velocity of certain particles cannot be pinpointed conjointly. This renders the question "What is the exact position and velocity of particle X at time t?" not insoluble but illegitimate. Question illegitimacy represents a limit that grows out of science itself—a limit on appropriate questions rather than on available solutions. Insolubilia, however, are something very different: they are legitimate questions to which no answer can possibly be given—now or ever.

Any claim to identify insolubilia by pinpointing here and now questions that science will never resolve is bound to be problematic—indeed, extremely far-fetched. The conception of identifiable insolubilia runs into deep theoretical difficulties. Charles S. Peirce has put the key point trenchantly:

> For my part, I cannot admit the proposition of Kant—that there are certain impassable bounds to human knowledge.... The history of science affords illustrations enough of the folly of saying that this, that, or the other can never be found out. Auguste Comte said that it was clearly impossible for man ever to learn anything of the chemical constitution of the fixed stars, but before his book had reached its readers the discovery which he had announced as impossible had been made. Legendre said of a certain proposition in the theory of numbers that, while it appeared to be true, it was most likely beyond the powers of the human mind to prove it; yet the next writer on the subject gave six independent demonstrations of the theorem.[19]

To identify an insolubile, we would have to show that a certain scientifically appropriate question is such that its resolution lies beyond every (possible or imaginable) state of future science. This is clearly a very tall order—particularly so in view of our inevitably deficient grasp on future states of science.

How could we possibly establish that a question Q will continue to be raisable and unanswerable in every future state of science, seeing that we cannot now circumscribe the changes that science might undergo in the future? We would have to argue that the answer to Q lies "in principle" beyond the reach of science. This would gravely compromise the legitimacy of the question as a genuinely scientific one. For if the question is such that its resolution lies in principle beyond the powers of science, it is difficult to see how we could maintain it to be an authentic scientific question.

The best we can do here and now is to put Q's resolvability beyond the power of any future state of science that looks to be a real possibility from where we stand. But we shall never be in a position to put it beyond the reach of possible future states of science as such. If a question belongs to science at all—if it reflects the sort of issue that science might possibly resolve in principle and in theory—then we cannot categorize it as an insolubile. With respect to science, we have no alternative to adopting the principle that *what can be done in theory might be done in the future.*

A dismissal of insolubilia might meet with the objection that present-day science seems to put the answers to certain (perfectly meaningful) factual questions wholly beyond our reach. For example, let the "big bang" theory of cosmic origination be taken as established, and suppose further that its fundamental equations take such a form that, owing to the singularities that arise at the starting point $t = 0$, no inference whatever could be drawn regarding the state of things on "the far side" of this starting point—the time preceding the world-originating big bang. Then a question like "Was the physical universe pretty much like ours before the big bang, or were the laws of physics of the preceding world-cycle different?" would *(ex hypothesi)* not be answered. Yet, this would be so not because this is an inherently disallowed (scientifically meaningless) question but because there could be no possible way of securing the needed data. Its status would be akin to that of questions about the mountains on the other side of the moon in Galileo's time.

Herein lies the crux. The question at issue is not necessarily insoluble as such: it is simply that the existing state of science affords no way of getting an answer. As we have seen, present science cannot speak for future science. Thus there can be no basis for claims of inherent unanswerability—no way of justifying the claim that an answer will remain unattainable in every future state of science. It has eventuated, amazingly enough, that the big bang itself has left traces that we ourselves can detect and use to make inferences about its nature. Who is to say that we may not one day discover, if we are sufficiently ingenious, that earlier cosmic cycles somehow leave a trace on their successors? It would be ill-advised to be so presumptuous as to say, on the basis of a mere few centuries of experience with science, that the problems that seem insoluble to us today are also destined to perplex our successors many thousands of years further down the corridors of time.

A key point to emerge from these deliberations is that, as regards questions, science does not have limits. In his vivid Victorian prose, Baden Powell put the matter well more than a century ago:

> It is the proper business of inductive science to analyse whatever comes before it. We cannot say that any physical subject proposed is incapable of such analysis, or not a proper subject for it, until it has been tried and found to fail; and even then, the result is not unprofitable, ... for the unknown regions on the frontier of science enjoy at least a twilight from its illumination, and are still brightened by the rays of present conjecture, and the hope of future discovery. We can never say that we have arrived at such a boundary as shall place an *impassable* limit to all future advance, provided the attempts at such advance be always made in a strictly inductive spirit. To the truly inductive natural philosopher, the notion of limit to inquiry is no more real than the mirage which seems to bound the edge of the desert, yet through which the traveler will continue his march to-morrow, as uninterruptedly as to-day over the plain.[20]

No adequate justification can be found for the view that science has barriers—that there are facts in its domain that science cannot in principle ascertain. (Something that is *in principle* unattainable by science could not justifiably be held to belong to its province at all.)

The inherent unpredictability of science has the immediate consequence that no relevant issue can securely be placed outside its reach. To put particular explanatory issues outside the effective capacity of science, we would need to develop an argument of roughly the following format:

(1) Scientifically acceptable explanations will always have such-and-such a character (are Z-conforming).

(2) No Z-conforming explanation will ever be devised to resolve a certain explanatory question.

(3) The explanatory question at issue must thus always remain outside the scope of scientific explicability.

Of course, this sort of argumentation accomplishes no more than to show the incompatibility of not-(3) with (1) and (2). Thus, very different approaches are possible. One could always accept (3) and deny either (1) or (2). Consider, in particular, the prospect of (2)-denial. The only way to show that no *future* state of science can ever resolve a certain question is to show that no *possible* state of science can resolve the question—that no possible state of science can bear one way or another toward answering it. At this stage, a dilemma comes into operation with respect to the sort of possibility at issue. For this can either be *epistemic* possibility (based on what we know) or *theoretical* possibility (based on abstract general principles). But argumentation with respect to the first sort of possibility founders on the fact that present science cannot speak for future science—that "what we (take ourselves to) know" in science cannot be projected into the future. Argumentation with respect to the second sort of possibility fails because to show this sort of impossibility would be to put the question outside the range of issues appropriately characterizable as "scientific." (Natural science is not rendered *incomplete* by its failure to deal with the subtleties of Shakespearean imagery!)

Moreover, one can always argue that (1) embodies an overly narrow and restrictive view of science. One could simply deny that science embodies the particular delimitation at issue in Z-conformity. The inherent plasticity of science means that this approach will always be available to us.

To be sure, someone might object as follows: "One can surely develop a plausible inductive argument on behalf of insolubilia along the following lines: we establish (1) that Q cannot be answered unless X is done, and (2) that there is good reason to think that X cannot be done at all—now or in the future." For example, it might be maintained that the question "What physical processes go on inside a black hole?" cannot possibly be answered unless we manage to secure hole-internal observations (that is, make them and get information about them out), and that this cannot be done because the hole exerts a "cosmic censorship" that imposes an insuperable barrier on data extraction. The vulnerable point here is the contention "Q cannot be answered unless X is done" (that hole-internal processes cannot be characterized unless we extract physical data from the hole). Who is to say how future science can resolve its questions—seeing that it need certainly not do so in ways circumscribed by our present conceptions? (Surely, for example, considerations of theory might in principle—and to some extent actually do in practice—enable us to describe "what goes on in a black hole.") We cannot ever cogently establish that answering a question hinges unavoidably on doing something impossible (for example, sending a signal faster than the speed of light). We can no more circumscribe how science will realize its future tasks than we can circumscribe what those tasks will be.

The course of wisdom thus lies with the stance that there are no particular scientific questions—and certainly no presently identifiable ones—that science cannot resolve as a matter of principle. If a question does indeed belong to the province of science (if it does not relate to belles lettres or philosophy or musicology, etc.), then we have to accept the prospect of a scientific answer to it. No scientifically appropriate question can plausibly be held to be beyond the powers of science by its very nature: there can be no reason to think that any such question lies beyond its reach as a matter of principle. For the long and short of it is that questions cannot at one and the same time qualify as authentic scientific questions *and* be such that their answers lie in principle beyond the reach of science.

The idea of identifiable insolubilia accordingly shipwrecks on the same rock as the idea of domain limitations—namely, on the essentially unpredictable nature of science in its substantive regards. The cardinal fact is simply that no one is in a position to delineate here and now what the science of the future can and cannot achieve. No identifiable issue can confidently be placed outside the limits of science.

9 The Price of an Ultimate Theory

1. The Principle of Sufficient Reason

How does nature work at its most fundamental level? What is the key to understanding the modus operandi of the physical universe? This is one of the biggest questions of them all. Some science theoreticians are prepared to think big. The dream of an ultimate theory that explains it all has enchanted philosophers and scientists throughout the centuries. In the form of a grand, all-encompassing, unified "theory of everything" it continues to beguile physicists in our own day, even long after professional philosophers have given up on it.

A glint in the eyes of many physicists nowadays is the vista of a single overarching theory that achieves a synoptic explanatory unification of the laws of operation of the fundamental forces at work in physical nature: gravity, electromagnetism, the weak and strong nuclear forces, and perhaps also a somewhat diffuse force of symmetry tropism.[1] Somewhat in the manner of a "superstring" theory that involves all known physical force under the aegis of minuscule stringlike vibrations in space—characterized by Steven Weinberg as "perhaps the first plausible candidate for a final theory"[2]—it should provide an integrating principle for the fundamental process of physics. What is at issue here is the ideal of a single and unified principle of explanatory understanding that is at once all-embracingly comprehensive and also definitive in representing the end of the explanatory line, standing, so to speak, at the top of the food chain of explanatory nourishment in natural science. Such a theory is to constitute the explanatory pivot for all the major questions with which theoretical physics (at least) has to deal. Its task is to provide a grand unified theory that serves as a universal engine for the accomplishment of explanatory work. The pivotal idea is that of securing a key to unlocking the cardinal secrets of nature so as to render physical reality comprehensively intelligible.

Considerations of general principle would tend to suggest that this idea of an ultimate scientific theory is bound to encounter large obstacles en route to practicable realization. Even before the substantive matters can be addressed profitably, however, a good many preliminary clarifications of the theoretical and philosophical background issues are in order.

Confronted by any significant fact, the human mind, by its evolution-imprinted nature, seeks to know "the reason why." To all appearances, we stand committed to the idea encapsulated in the classical *principium rationis sufficientis*, the "Principle of Sufficient Reason," to the effect that there is always some explanation for why things are as they are, some basis for reaching an understanding of why it is that what is so is so and what is not is not. Let us consider just what it is that this involves.

We shall here let the variables t, t', t'', etc. range over the set T of factual truths about the physical world.[3] Let us further introduce the abbreviation $t \Sigma t'$ to be construed as "t carries the main burden in providing an answer to the explanatory question: 'Why is it that t' obtains?' (or equivalently 'What is it that explains that t'?)." (Note that here t does not represent time.)

With an explanatory relationship of this general sort in hand, we can delineate the idea of a Principle of Sufficient Reason to the effect that every fact is capable of being explained: that any fact whatsoever can be fitted out with an adequate explanation. In its simplest and most general formulation, such a Principle of Sufficient Reason asserts:

(PSR) $(\forall t)(\exists t')(t' \Sigma t)$

As Leibniz put it long ago in section 32 of the *Monadology* (1716): "[A principle] of sufficient reason [obtains], in virtue of which we consider that no fact could be true or actual, and no proposition true, without there being a sufficient reason for its being so and not otherwise, although most often these reasons cannot be known by us." This "Principle of Sufficient Reason" affirms that all of our questions about the world and its ways have available answers (however difficult it may prove in practice to come by them). In effect, the principle asserts that *for every fact there exists a cogent explanation for its being exactly as it is*. To put it in Hegelian terms, the principle guarantees that the real is rational.

Of course all this still leaves open the matter of just what a satisfactory explanatory answer to a why question would involve. Science theorists have in recent years devoted much effort to clarifying this issue. But for present purposes we may suppose that it is sufficiently clear to enable people to recognize an acceptable explanation when they see it. As a first approximation we can resort to the familiar "Hempelian model of explanation."[4] On this approach, explanation proceeds as follows: if t is a fact to be explained (an *explanandum*) then we can secure some more fundamental (that is to say more general and encompassing) fact t' which, in conjunction with some further, subsidiary, situation-characterizing fact t'' serves as the explanatory *explanans* in that it inferentially entails t: $t'\Sigma t$ iff $(\exists t'')[c(t'') \& ([t' \& t''] \to t)]$.[5] Here $\to$ represents inferential derivation and c indicates that the pertinent fact belongs to the "boundary value conditions" for the issue under consideration—the conditions that delineate the particular situation at hand in the factual context within which the explanatory problem arises. The resulting picture of explanatory rationalization looks like figure 9.1.

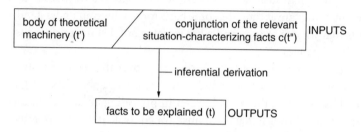

Figure 9.1   Hempelian Explanation

Here the explanatory process functions in such a way that we would explain facts with reference to theories and explain theories themselves, insofar as possible, with reference to other, broader, more high-level theories. Explanation power, like hydroelectric power, cascades downward to ever lower levels of theoretical generality.

On this basis, the Principle of Sufficient Reason takes the following format:

$$(\forall t)(\exists t')(\exists t'')[c(t'') \& ([t' \& t''] \to t)]$$

In effect, such a Principle of Sufficient Reason holds that with questions about the world's hows and whys we in principle need never confront a situation of total bafflement: whenever matters in the world stand X-wise (for any actual state of affairs whatsoever), then there is a rationale—an in-principle discoverable explanation—for why this is so.

With Immanuel Kant, who further elaborated the matter a century after Leibniz, the Principle of Sufficient Reason in its application to this spacetime world of ours took the form of a Principle of Causality. This principle asserts that everything that occurs in nature—every one of the world's eventuations—has a *causal* explanation: every occurrence is the effect of some cause from which it proceeds under the aegis of the causal laws governing nature's modus operandi. Throughout nature, it is efficient causation that always provides for the reason of things. Not only can the world's facts always be explained, but they can always be explained in terms of specifically causal laws.

In the heyday of Newtonian physics this principle was a definitive axiom of scientific philosophy. However, with the rise to prominence of stochastic phenomena in physics and the accompanying emergence of statistical laws, this classically deterministic picture of a world operating under strict causal laws went into decline. The Principle of Causality has vanished in its wake. Probability, chance, and indeterminism come to the fore.

Even though the Kantian Principle of Causality ultimately collapsed, the Principle of Sufficient Reason itself survived this blow. The idea of universal explicability under the aegis of natural laws kept its hold even as these laws changed from a deterministic to a probabilistic character. In place of the theory of a universally operative causality something rather different emerged.

Instead of the idea of a set of causal laws that can explain anything that occurs in nature, contemporary physics has been drawn to the idea of a grand unified theory (GUT)—a single diversified theory that, even if it cannot explain everything, can at any rate explain everything that is explicable. This conception of an ultimate theory has emerged as the modern substitute for the unrealizable old-style Principle of Causality. For while it does not say that everything can be explained on *causal* principles it says that everything can be explained on GUT-provided principles.

Exactly what is it that would be required for an ultimate explanatory theory of the sort envisioned by GUT enthusiasts? Somewhat surprisingly, physicists tend not to be very explicit on this critical background issue. Let us take a closer look at it from a philosophical point of view.

2. The Idea of an Ultimate Theory

An ultimate theory is not, of course, claimed to be *chronologically* ultimate in the sense of temporal finality—of averting any prospect of change and improvement in the wake of scientific progress. Rather, the theory must be final in the very different sense of answering all of the substantial questions we ask about the modus operandi of nature. It need not be final in the sense of being frozen in time and admitting no further improvement and refinements—of being totally impervious to the prospect of improvement in the light of further information. It must do its synoptic explanatory work but need not be exempt from the prospect of a further development that enables it to do so more effectively and efficiently. Its purported ultimacy is strictly explanatory and thus *hermeneutic* rather than *chronological*. But just what does this involve?

An ultimate theory must be holistic: it must have the integrity of a collective whole. It cannot be a distributive collective of parts and pieces but must indeed be a grand unified theory. For we must reject the philosophical doctrine of what has come to be called the Hume-Edwards thesis that: "if the existence of every member of a set is explained, then the existence of the set is thereby

explained."⁶ This thesis is predicated on a fallacy in a way that can be seen as follows. Consider the following two claims:
- If the existence of every member of a team is explained, the existence of that team is thereby explained.
- If the existence of each member of a criminal gang is explained, the existence of that criminal gang is thereby explained.

Both of these claims are clearly false as they stand. On the other hand, contrast these two with the following cognate revisions:
- If the existence of every member of a team *as a member of that particular team* is explained, then the existence of that team is thereby explained.
- If the existence of every member of a criminal gang *as a member of that particular criminal gang* is explained, then the existence of that criminal gang is thereby explained.

Both these theses are indeed true—albeit only subject to that added qualification. After all, to explain the existence of the bricks is not automatically to achieve an explanation of the wall, seeing that this would call not just for explaining these bricks distributively but their collectively coordinated co-presence in the structure at issue.

And the case is just the same with the Hume-Edwards thesis:
- If the existence of the members of a set is explained, the existence of that set is thereby explained.

This too is acceptable only if construed as:
- If the existence of every member of a set *as a member of that particular set* is explained, then the existence of that set is thereby explained.

Now if this condition of affairs indeed obtains, then the Hume-Edwards doctrine fails. For to get a distributive explanation of *the right sort* in point of viability we will need a collective explanation of just the sort that those distributionists are seeking to avoid.

The fact is that the adherents of the Hume-Edwards approach are caught on the horns of a dilemma. They can have their pivotal thesis either as it stands or as duly qualified with reference to a required commonality. In the former case, the thesis is false. In the latter case it is, although true, unable to bear the reductive burden they wish to place upon it.⁷

The lesson of their deliberations is that ultimate theory has to be duly holistic. What is needed for a synoptic explanation—and what a GUT theory obviously contemplates—is a specific, integral theory able to achieve the explanatory task on a collective rather than distributive basis. The pathway to an ultimate theory will therefore not be an easy one.

The unity of science to which many theorists aspire may indeed come to be realized at the level of concepts and theories shared between different sciences—that is, at the level of ideational overlaps. But for every conceptual commonality and shared element there will emerge a dozen differentiations. The increasing complexity of our world picture is a striking phenomenon throughout the development of modern science. It is, of course, possible that the development of physics may eventually carry us to theoretical unification where everything that we class among the laws of nature belongs to one grand unified theory—one all-encompassing deductive systematization integrated even more tightly than that of Newton's *Principia Mathematica*. But the integration at issue at the pinnacle of the pyramid will cover further down an endlessly expansive range and encompass the most variegated components. It will be an abstract unity that succeeds in uniting a concrete manifold of incredible variety and diversity.

Thus the upshot on what is clearly the most direct and natural construction of the ultimacy that a "theory of everything" is supposed to exhibit, two features are principally involved in such an "ultimate" or "final" theory that is internally unified and externally synoptic—let us designate it by T^*:

- Explanatory *Comprehensiveness* [C]
Wherever there is a fact, T^* affords its explanation:

$$(\forall t)(T^* \Sigma t)$$

In effect, such explanatory comprehensiveness stipulates erotetic completeness with respect to explanatory questions—about the modus operandi of nature in particular.

It should be noted that this condition at once entails $(\exists t')(\forall t)(t' \Sigma t)$. This in turn is something that entails—and is in fact significantly stronger than—the mere Principle of Sufficient Reason: $(\forall t)(\exists t')(t' \Sigma t)$.[8]

The second aspect of GUT-style theory is

- Explanatory *Finality* [F]
There is no further, deeper explanation of T^* itself:

$$\sim(\exists t)(t \Sigma T^* \,\&\, t \neq T^*) \text{ or } (\forall t)(t \Sigma T^* \rightarrow t = T^*)$$

This of course means that the possibilities are reduced to the point where the only conceivably appropriate explanation of T^* is T^* itself. With the possible exception of T^* itself, there is nothing else available that is in a position to provide an explanation of T^*: it stands at the end of the explanatory line.

The availability of a comprehensive and final explanatory theory along these lines—one that is capable of accounting for the laws of nature themselves—constitutes the core of the doctrine of grand unified explicability. As the very name suggests, unity and comprehensiveness must characterize any such "grand unified theory."[9]

3. An Aporetic Situation

Now the crucial consideration for present purposes is that the idea of an ultimate theory along these lines stands in decided conflict with a principle that lies at the heart of the traditional conception of explanatory adequacy, namely a stipulation to the following effect:

- *Explanatory Noncircularity* [N]

    No satisfactory explanation can invoke the very fact that is itself to be explained:

    $\sim(\exists t)(t \Sigma t)$

This clearly fundamental noncircularity principle obtains because it is obviously problematic from a probative point of view to deploy a theory for its own explanation. No blatantly circular explanation can ever be altogether satisfactory, seeing that it presumes the explanatory availability of the very item whose explanation is at issue.

Here is the point where trouble ensues for any sort of "ultimate" theory. For observe that Comprehensiveness at once entails $T^* \Sigma T^*$. Moreover, there is another route to the same destination even without proceeding via comprehensiveness:

1. $(\exists t) t \Sigma T^*$ by PSR
2. $T^* \Sigma T^*$ from 1 by Finality

Both of these pathways lead us to a conflict with Noncircularity, so that both Comprehensiveness and Finality are in trouble as long as Noncircularity is maintained.

Diagramatically we have the following situation with respect to implication relationships:

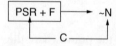

Figure 9.2 The Interrelation of Principles

What this means is that in espousing *N* one must abandon *C* and (if one wishes to retain *PSR*) also *F* as well. And so a GUT theorist committed to *C* and *F* is bound to regard *N* as a principle that has to be jettisoned.

Noncircularity is not so easily abandoned. For what we now have is the distinctly problematic idea of circular self-explanation at exactly the most crucial juncture of our explanatory project. The unhappy precedent of *causa sui* theorizing in theology looms before us here. This problematic option exacts what is, to all appearances, an altogether unacceptable price.

In the background of this perplex lies the fact that the fatal flaw of any purported explanatory "theory of everything" arises in connection with the ancient paradox of reflectivity and self-substantiation. How can any theory adequately substantiate itself? *Quis custodiet ipsos custodes*? What are we to make of the man—or the doctrine—that claims "I stand ready to vouch for myself"? How can such self-substantiation possibly be made effective? All of the old difficulties of reflexivity and self-reference come to the fore here.[10] No painter can paint a comprehensive picture of a setting that includes this picture itself. And no more, it would seem, can a theorist propound an explanatory account of nature that claims to account satisfactorily for that account itself. For insofar as that account draws on itself, this very circumstance undermines its viability.

The upshot of such considerations is that an adequate "ultimate theory" of scientific explanation, naively understood as satisfying all of the traditional explanatory desiderata, is in principle impossible.

The reason for this state of affairs is easily grasped. For consider again:

- (PSR) Every fact has a satisfactory explanation.
- (N) Nothing is self-explanatory: a satisfying explanation must always lie in something yet deeper.

These principles launch us into an explanatory regress that looks to the prospect of ongoingly deepening our explanation. And this regress—this seemingly indispensable escape route—is cut off by supposing the explanatively ultimate cul-de-sac of an ultimate explanation.

4. A Way Out of the Impasse

How can one find an exit from this labyrinth?

The most promising way out of the impasse seems to lie in going back to square one and taking another look at the very idea of explanation itself. This being so, a GUT theorist is well advised to go back to the old drawing board and revamp the naive theory of explanatory adequacy.

As noted above, contemporary explanation theory generally construes this on the lines of the deductive model urged in the classic pages of Hempel and Oppenheim. Recall the tenor of the Hempelian position regarding the modus operandi of explanation, which can be represented as follows:

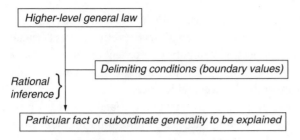

Figure 9.3 Hempelian Explanation

Given this understanding of the matter, explanation clearly becomes something that has to move from the more basic to the derivative. On Hempelian principles, the explanatory process continues to implement Aristotle's idea of substantive priority that sees the *explanans* as more fundamental and far-reaching than the *explanandum*. Explanation stands committed to unidirectional priority.

On this basis there would be no reason to countenance the prospect of an ultimate theory. As Karl Popper put it in rejecting "the idea of an ultimate explanation," we must acknowledge that "every explanation may be further explained by a theory or conjecture of a higher degree of universality. There can be no explanation which is not in need of a further explanation."[11]

This idea of an endless generality regress is deeply problematic. In a universe based on the operation of fundamental forces that are, presumably, finite in number, it would indeed seem that an explanatory regress will have to terminate in some altogether unexplained explainer. We seem to be driven to something that is supposed to be explanatorily fundamental and yet is itself altogether unexplained.

There is, however, a promising alternative here—a very different way of understanding how explanation can work. For instead of giving our explanations through an inferential subsumption under prior, more fundamental theses, one can take resort to a variant mode of explanation—one that proceeds not via prior premises but via posterior consequences. Such an approach offers the option of explaining a fact through the consideration that it itself serves to account for a significant spectrum of other things. On this basis our explanations proceed not by derivation from basics but through the fertility of consequences—and thus through their systemic role in the overall fabric of inferential systematization. We are thus led to a dualized conception of explanation as a process that can proceed either by way of direct *derivation* (as per →) or by way of the *systematization S* that is achieved when t is inferentially embedded in the wider setting of an overall system of explanatory understanding. In thus going to a two-tier mode of explanation, we envision two different ways of "explaining" a thesis t: (1) *inferentially* when $(\exists t')(t'\Sigma t)$, and (2) *systemically* when $S(t)$.

On such an approach we explain facts by accounting for them in terms of their lawful connections with others and continue this process as long as necessary. But how are we to explain the laws of nature themselves (or purported laws of nature, strictly speaking)? Will law explanations not always leave something unaccounted for?

In general, laws are explained by their derivation from other laws. This is clearly a process that must come to a stop. The set of laws of nature—as best we can develop it—is always finite. With facts we can in principle go on *ad indefinitum* in referring them to others, but with laws this process must have a stop: the subsumptive explanation of laws must eventually come to the end of the road. Here we will always reach certain ultimate, axiomatic principles. How, then, do we explain these?

Do we not at this stage reach a limit of sorts? If the ultimate or fundamental laws play the role of basic premises in science, much as the axioms are basic in a system of geometry, does this not render them inexplicable? For although science uses them in giving explanations, they will themselves lie outside the range of scientific explicability.

The answer here lies in recognizing that the axioms of our law systems are not inexplicable ultimates. For one thing, their being axiomatic is largely incidental. Axioms, after all, are axioms not because of what they say but because of how we fit them into the organization of our knowledge. Axiomaticity is a matter of the particular systematization of information we find it convenient to adopt. (What is axiomatic in one system might well be theorematic in another.) Our axioms reflect how we think about the workings of nature rather than how nature itself is constituted. Aristotle notwithstanding, the relative fundamentality of laws and facts regarding nature is something that lies in the eyes of the beholder and not in the nature of things.

Moreover, axioms too are explicable. Not, to be sure, through derivation from other, still more fundamental theses—which would, after all, make *these* the true axioms, and the original ones mere pretenders. Rather, axioms find their explanation *systemically*, in terms of their role in the whole system—as being needed to give rise to the whole elegantly and economically. Even so must "ultimate" laws be accounted for in terms of their integrative role in the entire law system.

The entire subsumptive approach stands in contrast to a very different one that proceeds *nonsubsumptively* on the basis of "best-fit" considerations. The best fit ("coordination") at issue may involve *inferential connectedness*, but it may also be a matter of analogy, uniformity, simplicity, and the other facets of systematization. Thus, we have here a wholly different approach to explanation; one that takes systematization itself as the key, relying not on subsumptive inference but on systemic coordination. Here the issue is not *subordination* to but *coordination* with: it is not a matter of *inference* from other theses at all but one of optimal meshing through mutual attunement with them. On such an approach, if we explain A with reference to B and C, we do so not by inferring A from B and C but by showing that A is more smoothly co-systematizable with B and C than is the case with any of its alternatives, A', A'', and so on.

The idea of an explanatory bedrock—of certain ultimate, axiomatic, and inexplicable theses that simply do not need explanation—is not very satisfying and is, moreover, quite unnecessary. The sensible view is simply that different modes of explanation are at issue at different levels of discussion. In endowing the theories of science with an axiomatic development, we "explain" the theorems *subsumptively* because the axioms yield them, but we explain the axioms *coordinatively* because they engender the appropriate theorems and provide for the right systemic results. The law axioms are explained teleologically, as it were, because the processes they represent must obtain in order for the system to function as it does. The "ultimate laws" are explained through their capacity to underwrite the holistic integrity of the entire system.

Accordingly, axioms are not inherently unjustifiable. Systemic unity serves as the overarching principle of explanatory legitimation in providing the framework through which axioms themselves are accounted.[12]

In the wake of this enlarged understanding of explanation we now change the specific nature of comprehensiveness from $C = (\forall t)[T^*\Sigma t]$ to: (C') $(\forall t)[T^* \Sigma\, t \vee S(t)]$. Moreover, we correspondingly change the Principle of Sufficient Reason (PSR) from $(\forall t)(\exists t')(t'\Sigma t)$ to: (PSR') $(\forall t)[(\exists t')(t'\Sigma t) \vee S(t)]$. It is now clear that these revised versions of C' and PSR' are perfectly compatible with Finality (F) and Noncircularity (N).

The perplex of aporetic inconsistency that initially confronted us is now avoided: there is no longer any sort of conflict among our various fundamental theses. The revised principles just conspire in indicating that in the particular case of T^*, at least, the proper mode of explanation lies in the S-mode, so that we have $S(T^*)$. In taking recourse to a second mode of systemic explanation represented by S and readjusting our theses in its light we thus restore peace and harmony in the family of explanatory principles.

Moreover, it seems in any case to be desirable to effect this change in the construction of PSR. For as long as we interpret this principle with *explanation* understood in a subsumptive (Hempelian) manner we cannot avert the Hobson's choice between circular explanation or an infinite regress—as Aristotle already indicated in his *Posterior Analytics*. So the recourse to explanatory systematicity seems to be desirable in any case.

How is the sort of systematization at issue here to be understood: how does S work? Specifically, what will systematization look like at the level of an ultimate theory T^* at the cosmological level? Recall here the three basic statements of our account:

T^* = the "grand ultimate theory" itself.

$T = t_1 \& t_2 \& \ldots \& t_n$ = the conjunction of our various detailed theories of physical explanation.

$B = b_1 \& b_2 \& \ldots \& b_n$ = the conjunction of the basic observed and calculated boundary value conditions for physical explanation—including, in particular, the fundamental constants of nature.

In looking to systematization at this level we would, presumably, have three desiderata in view:

1. $T^* + B$ "explains" T—and does so in something like the Hempelian subsumptive manner. The conjoining of T^* with the various boundary-value conditions should provide a suitable (subsumptive) explanation of principal physical theories themselves.

2. $T + B$ "evidentiates" T^*. T^* should be an optimal solution to its question: given B and T, what sort of unified higher-level account provides for the optimal systemic unification of the components of T?

3. $T^* + T$ "determines" B. Insofar as possible, the purely theoretical considerations of T^* and T should constrain the fundamental constraints of nature and the world's fundamental boundary nature conditions. But where such value-constraint is unachievable, it should at least transpire that the theoretical resources render their actually observed values maximally probable, so operating that (in the manner of the Anthropic Principle) they render the determinable boundary-value conditions of nature maxiprobable in relation to possible alternatives. The characteristic features of the world as we have it should be probabilized by the laws of nature that we accept.

In this way we have the situation that the trio T^*, T, and B is so interdependently connected that, given any two of its members, the third becomes fixed in place. This sort of integrating interconnection is, clearly, the optimal way of providing for the systematic integration of our ultimate theory T^* within the wider framework of our physical knowledge. And it would, ideally, be in its capacity to play this integrative role that the systemic "explanation" of our ultimate theory consists.

5. Implications

These considerations put us into a position to deal with the holistic question of why the overall framework of scientific fact and law and system should be as it is. If explanation at the level of facts and laws is inevitably intrasystemic, what is to explain the whole system itself?

In a way, this question is an invitation to folly. We clearly cannot provide a scientific explanation for the whole system of science in terms of something that falls outside: it would not be "the whole system" if anything fell outside it. Explanatory self-subsumption is infeasible at the level of facts and laws, but at the systemic level it is a conditional necessity: if the system can be explained at all, that explanation must fall within it. We cannot get outside the framework of our completed explanatory system: to explain the system in terms of X would simply be to enlarge it to include X itself. The quest for a system-external foundation for the scientific rationalization of the system is ultimately senseless.

We explain facts in terms of facts and laws. We explain laws in terms of facts, laws, and systemic principles. And we explain systems self-referentially in terms of the laws they are able to systematize. The laws that comprise the system must have a substantive content that renders certain structural systemic principles (for example, universality, economy, homogeneity) paramount, principles that, in their turn, serve to account for the character of those laws themselves. The laws, that is, stipulate design features of such a sort that a law system optimally designed in line with these features will have exactly the character of the laws at issue.[13]

The resultant situation is indeed cyclical, but this simply reflects the structural coherence of rational systematization. Facts are explained from above, via the laws they instantiate; systems from below, via the laws they rationalize. Laws can be approached from either end—via the facts they coordinate or via the family of laws with which they stand in systemic coordination. The overall process is not a vicious but a virtuous circle of self-substantiation. Adequacy lies not in a unidirectional flow from the more basic to the less so but in the smooth meshing of the overall cycle.[14]

Our principles of systematization consequently have the feature that they are themselves monitored by an overarching structure of systemic order. This closing of the cycle of substantiation is itself an aspect of systematicity. The adequacy of our explanatory systematizing is thus itself controlled by systematic considerations. In the final analysis, then, we explain the system-as-a-whole through its capacity to "do the job" of scientific rationalization. This, however, is not a defect of our explanatory capacities but an immediate and inevitable consequence of the nature of an explanatory system.

The issue of legitimation is thus settled in terms of a cyclic interdependence and self-supportiveness. The idea of explanatory stratification is misleading: no neat linear order of fundamentality obtains among nature's facts or laws. No useful work can be done for us by the Aristotelian principle that explanation must always and everywhere proceed unidirectionally from the less to the more rudimentary (or basic, or what have you). The ancient search for linear "priority" in matters of scientific explanation can be carried only so far.

As this perspective indicates, envisioning a GUT theory calls for accepting a significant complication. Specifically, we must complicate our concept of explanation through revising our understanding of how theories are explained by taking a two-track approach to explanation either via antecedents or alternatively via consequences. Over and above the prospect of explanation by inference from more fundamental priors we must resort to the idea of "explaining" a theory in terms of its posterior consequences—that is, in terms of what goes after rather than what came before. We must be prepared on occasion to resolve our explanatory question by nonstandard means and to adopt—at least in some cases—a *coherence theory* of explanation. For in at least one case—that of the ultimate theory itself—we must "explain" matters in the same—and indeed the only—way in which we can explain the axioms of a formalized theory: not by derivation from something yet more fundamental, but by taking note of the circumstance that it represents a commitment that leads to the right sorts of consequential results.

We can return to the issue posed by our initial question of the cost of an ultimate theory. This, so it now emerges, is that we must modify—and complicate—the generally accepted view of the nature of explanation itself. For we are constrained to adopt a more sophisticated account of explanation—a two-track conception, proceeding either by way of derivation from more fundamental antecedents or by a systemic role that involves fertility and centrality in the demonstration of consequences.[15]

In theory, to be sure, we need only resort to this complication of dualized explanation in the case of the ultimate theory itself. There is, however, no need to be quite that squeamish about this useful device. For what now clearly emerges is that in *some* cases, at any rate, it deserves to be seen as natural and appropriate to take such a perspectively-oriented approach. This is something that is in principle capable of extension across a wider range.

The fact of the matter is that such a complication in our concert of explanation brings with it some not insignificant theoretical advantages. For what we now have is, in effect, a dualizing of explanatory dimensions. An explanation can now be extended in two different ways. Either in *depth,* by furthering the regressive search for reasons why behind the reasons why, or in *breadth,* by extending the patterns of systemic coherence through which we amplify the systemic enmeshment of the item to be explained into the wider framework of our relevant commitments. What is lost in simplicity in dualizing our concept of explanation is more than compensated for by what we gain in range and power.

Viewed as a theoretical desideratum, an "ultimate" theory is to be a unified explanatory principle that dispenses with a disaggregated proliferation of explanatory devices. It enables us to discharge the explanatory mission of physical science on a unitized and collective rather than a fragmented and distributive basis.

However, as the preceding deliberations indicate, the question of how such a "final" theory itself is to be explained cannot be avoided. It is here that a compensatory price is exacted from us. For we have no alternative but to see the explanation of an ultimate theory to lie in the diversified justification of explanatory work that it facilitates. At this point diversity and plurality force their way to the forefront again. For the explanatory justification of the theory itself is something that is achieved only distributively on the basis of the complex array of explanatory tasks that the theory is able to accomplish. On the revised explanatory perspective that is now at issue it is the very considerations that qualify a theory to be considered as a "grand unified theory" that themselves now serve to "explain" that theory and thereby to provide its legitimative substantiation.

The moral of the story, then, is that the price of an explanatory program that envisions a "first principle" of some sort is having to accept a more complex and sophisticated two-track approach to the business of explanation. If "ultimate" questions are to be resolved satisfactorily, what is needed is not so much an ultimate *answer* as an ultimate *process* for the provision of answers, a process for which cognitive systematization provides the cornerstone. On this basis the "explanation" of our explanatorily "ultimate" or "final" theory itself lies in the very fact of its being able to accomplish effectively and efficiently the synoptic explanatory task for which it is designed. The price of an ultimate theory, then, is thoroughgoing systematization.

6. Historical Postscript

In closing, a brief historical postscript is in order. For the basic idea that is at work in the preceding account goes back to the very dawn of speculative thought about the nature of explanation—to Plato's discussion in the *Republic* (at book 7, 510 B.C.).

> In studying geometric matters, the mind is compelled to employ assumptions, and, because it cannot rise above these, does not travel upwards to a first principle; and moreover the mind here uses diagrams as images of those actual things. However, this mathematical sector contrasts with the [philosophical] sector of the intelligible world which unaided reasoning apprehends by the power of dialectic, when it treats its assumptions, not as first principles, but as *hypotheses* in the literal sense, things "laid down" like a flight of steps up which it may mount all the way to something that is not hypothetical, the first prin-

ciple of all. Then, having grasped this, the mind may turn back and, holding on to the consequences which depend upon it, descend at last to a conclusion, never making use of any sensible object, but only of Forms, moving through Forms from one to another, and ending with Forms. [And so we] distinguished the field of intelligible reality studied by dialectic as having a greater certainty and truth than the subject-matter of the "arts," as they are called, which treat their assumptions as first principles. The students of these arts are, it is true, compelled to exercise thought in contemplating objects which the senses cannot perceive; but because they start from assumptions without going back to a first principle, you do not regard them as gaining true understanding about those objects, although the objects themselves, when connected with a first principle, are intelligible.

As such deliberations indicate, Plato too found the idea of unexplained explainers unpalatable. His complaint regarding Euclidean-style geometry, for example, is just exactly this—that it proceeds from first principles that are laid down as arbitrary stipulations ("absolute hypotheses") and not themselves fitted out with an explanatory rationale. By contrast, the great merit of philosophy—as he saw it—is that it treats its first principles not as absolute but as provisional hypotheses and that it proceeds not *deductively* but *dialectically*, looking down along the chain of consequences in order to substantiate the principle from which they derived their credibility.

Accordingly, Plato's position stressed the idea that for thoroughgoing rationality one must take philosophy's deeper dialectical approach of justifying one's beliefs cyclically, so to speak, by looking first upward to first principles, which themselves are then justified downward with reference to these consequences and ramifications. As Plato thus saw it, the standard process of mathematical justification of terms of absolute hypotheses that themselves remain unjustified—however customary in geometry or arithmetic—is not ultimately satisfactory from a rational point of view because it leaves off at the point where a different, dialectical methodology is called for.

Essentially this same line of reasoning is at issue in the two-tier conception of explanation to which—if the present account is anything like correct—we are driven in the course of trying to make workable sense of the conception of an ultimate theory in physical explanation. For the only really satisfactory validation of any purportedly ultimate commitment is one that invokes the overall performance of that commitment within the entire system with reference within which its ultimacy obtains.

The process of systematization that validates those seemingly axiomatic starting points also envisions something ultimate. What is ultimate here does not lie in the range of axioms, theses, or propositions but rather is something methodological—the "dialectical" process, as it were, by which such purported starting points become validated through cyclic and retrospective considerations. It is, in sum, not the axioms as such, but rather the process of their legitimation through those essentially retrospective systematic considerations that become available once they have done their work, that is properly to be seen as the rational crux of the matter.

# The Theoretical Unrealizability of Perfected Science

Synopsis

(1) Perfected science would have to realize four theoretical desiderata: (a) erotetic completeness (including explanatory completeness), (b) pragmatic completeness, (c) predictive completeness, and (d) temporal finality (the w-condition). (2-5) There is, as a matter of fundamental general principle, no practicable way for us to establish that any one of these desiderata is realized. Our science must be seen as inherently incompletable, with an ever-receding horizon separating where we are from where we would ideally like to be. (6) Perfection is dispensable as a goal for natural science; we need not presuppose its potential attainability to validate the enterprise. The motive force of scientific progress is not the a fronte pull of an unattainable perfection but the a tergo push of recognized shortcomings. (7) Perfected science is not a realizable condition of things but an idealization that provides a useful contrast-conception to highlight the limited character of what we do and can attain. (8) The unachievability of perfected science constrains us to recognize that natural science affords us no more than an imperfect picture of reality. The doctrine of "scientific realism" is thus unacceptable.

1. Conditions of Perfected Science

How far can the scientific enterprise advance toward a definitive understanding of reality? Might science attain a point of recognizable completion? Is the achievement of perfected science a genuine possibility, even in theory when all of the "merely practical" obstacles are put aside as somehow incidental?

What would *perfected science* be like? What sort of standards would it have to meet? Clearly, it would have to complete in full the discharge of natural science's mandate or mission. Now, the goal structure of scientific inquiry covers a good deal of ground. It is diversified and complex, spreading across both the cognitive/theoretical and active/practical sectors. It encompasses the traditional quartet of description, explanation, prediction, and control, in line with the following picture:

Table 10.1. The Goals of Science.

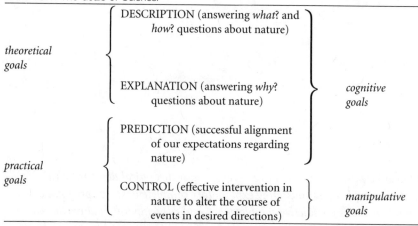

The theoretical sector concerns itself with matters of characterizing, explaining, accounting for, and rendering intelligible—with purely intellectual and informative issues, in short. By contrast, the practical sector is concerned with guiding actions, canalizing expectations, and, in general, with achieving the control over our environment that is required for the satisfactory conduct of our affairs. The former sector thus deals with what science enables us to *say*, and the latter with what it enables us to *do*. The one relates to our role as spectators of nature, the other to our role as active participants.

It thus appears that if we are to claim that our science has attained a perfected condition, it would have to satisfy (at least) the four following conditions:

1. *Erotetic completeness*: It must answer, in principle at any rate, all those descriptive and explanatory questions that it itself countenances as legitimately raisable, and it must accordingly explain everything it deems explicable.

2. *Predictive completeness*: It must provide the cognitive basis for accurately predicting those eventuations that are in principle predictable (that is, those that it itself recognizes as such).

3. *Pragmatic completeness*: It must provide the requisite cognitive means for doing whatever is feasible for beings like ourselves to do in the circumstances in which we labor.

4. *Temporal finality* (the omega condition): It must leave no room for expecting further substantial changes that destabilize the existing state of scientific knowledge.

Each of these modes of substantive completeness deserves detailed consideration. First, however, one brief preliminary remark. It is clear that any condition of science that might qualify as "perfected" would have to meet certain formal requirements of systemic unity. If, for example, there are different routes to one and the same question (for instance, if both astronomy and geology can inform us about the age of the earth), then these answers will certainly have to be consistent. Perfected science will have to meet certain requirements of structural systematicity in the manner of its articulation: it must be coherent, consistent, consonant, uniform, harmonious, and so on. Such requirements represent purely formal cognitive demands upon the architectonic of articulation of a body of science that could lay any claim to perfection. Interesting and important although they are, we shall not, however, trouble about these *formal* requirements here, our present concern being with various *substantive* issues.[1]

2. Theoretical Adequacy: Issues of Erotetic Completeness

Erotetic completeness is an unattainable mirage. We can never exhaust the possibility of questions. The Kantian Principle of question propagation means that inquiry—the dialectic of questions and answers—can never get to the ultimate bottom of things.

Any adequate theory of inquiry must recognize that the ongoing process of science is a process of *conceptual* innovation that always leaves certain theses wholly outside the cognitive range of the inquirers of any particular period. This means that there will always be facts (or plausible candidate facts) about a thing that we do not *know* because we cannot even conceive of them. For to grasp such a fact calls for taking a perspective of consideration that we simply do not have, since the state of knowledge (or purported knowledge) is not yet advanced to a point at which its entertainment is feasible. In bringing conceptual innovation about, cognitive progress makes it possible to consider new possibilities that were heretofore conceptually inaccessible.

The language of emergence can perhaps be deployed profitably to make the point. What is at issue, however, is not an emergence of *the features of things* but an emergence in our *knowledge* about them. Blood circulated in the human body well before Harvey; uranium-containing substances were radioactive before Becquerel. The emergence at issue relates to our cognitive mechanisms of conceptualization, not to the objects of our consideration in and of themselves. Real-world objects are conceived of as antecedent to any cognitive interaction—as being there right along, or "pregiven," as Edmund Husserl puts it. Any cognitive changes or innovations are to be conceptualized as something that occurs on our side of the cognitive transaction, and not on the side of the objects with which we deal.[2]

The prospect of change can never be dismissed in this domain. The properties of a thing are literally open-ended: we can always discover more of them. Even if we view the world as inherently finitistic and espouse a Principle of Limited Variety that has it that nature can be portrayed descriptively with the materials of a finite taxonomic scheme, there can be no a priori guarantee that the progress of science will not engender an unending sequence of changes of mind regarding this finite register of descriptive materials. This conforms exactly to our expectation in these matters. For where the real things of the world are concerned, we not only expect to learn more about them in the course of scientific inquiry, *we expect to have to change our mind about their nature and mode of comportment*. Be it elm trees, or volcanoes, or quarks that are at issue, we have every expectation that in the course of future scientific progress people will come to think differently about them than we ourselves do at this juncture.

Cognitive inexhaustibility thus emerges as a definitive feature of our conception of a real thing. In claiming knowledge about such things, we are always aware that the object transcends what we know about it—that yet further and different facts concerning it can always come to light, and that all that we *do* say about it does not exhaust all that *can* be said about it.

The world's furnishings are cognitively opaque; we cannot see to the bottom of them. Knowledge can become more extensive without thereby becoming more complete. This view of the situation is supported rather than impeded if we abandon a cumulativist/preservationist view of knowledge or purported knowledge for the view that new discoveries need not supplement but can displace old ones.

The concept of a thing so functions in our conceptual scheme that things are thought of as having an identity, a nature, and a mode of comportment wholly indifferent to the cognitive state of the art regarding them, and presumably very different from our conceptions of the matter. This is something we presume or postulate; it is certainly not something we have discovered—or ever could discover. We are not—and will never be—in a position to evade or abolish the contrast between "things as we think them to be" and "things as they actually and truly are." Their susceptibility to further elaborative detail—and to changes of mind regarding this further detail—is built into our very conception of a "real thing." To be a real thing is to be something regarding which we can always in principle acquire more information.

Much the same story holds when our concern is not with things but with *types* of things. To say that something is copper (or is magnetic) is to say more than that it has the properties we associate with copper (or magnetic things), and indeed is to say more than that it meets our test conditions for being copper (or being magnetic). It is to say that this thing *is* copper (or magnetic). This is an issue regarding which we are prepared at least to contemplate the prospect that we've got it wrong.

There is thus good reason on general principle to think that erotetic completeness is unachievable. But another line of consideration is no less decisive for our present purposes.

Could we ever actually achieve erotetic completeness (Q-completeness)—the condition of being able to resolve, in principle, all of our (legitimately posable) questions about the world? Could we ever find ourselves in this position?[3]

In theory, yes. A body of science certainly could be such as to provide answers to all those questions it allows to arise. But just how meaningful would this mode of completeness be?

It is sobering to realize that the erotetic completeness of a state of science S does not necessarily betoken its comprehensiveness or sufficiency. It might reflect the paucity of the range of questions we are prepared to contemplate—a deficiency of imagination, so to speak. When the range of our knowledge is sufficiently restricted, then its Q-completeness will merely reflect this impoverishment rather than its intrinsic adequacy. Conceivably, if improbably, science might reach a purely fortuitous equilibrium between problems and solutions. It could eventually be "completed" in the narrow erotetic sense—providing an answer to every question one can also ask in the then-existing (albeit still imperfect) state of knowledge—without thereby being completed in the larger sense of answering the questions that would arise if only one could probe nature just a bit more deeply. Thus our corpus of scientific knowledge could be erotetically complete and yet fundamentally inadequate. Thus, even if realized, this erotetic mode of completeness would not be particularly meaningful. (To be sure, this discussion proceeds at the level of supposition contrary to fact. The exfoliation of new questions from old in the course of scientific inquiry that is at issue in Kant's Principle of Question Propagation spells the infeasibility of ever attaining Q-completeness.)

The preceding considerations illustrate a more general circumstance. Any claim to the realization of a theoretically complete science of physics would be one that affords "a complete, consistent, and unified theory of physical interaction that would describe all possible observations."[4] But to check that the state of physics on hand actually meets this condition, we would need to know exactly what physical interactions are indeed possible. And to warrant us in using the state of physics on hand as a basis for answering this question, we would already have to be assured that its view of the possibilities is correct—

and thus already have preestablished its completeness. The idea of a consolidated erotetic completeness shipwrecks on the infeasibility of finding a meaningful way to monitor its attainment.

After all, any judgment we can make about the laws of nature—any model we can contrive regarding how things work in the world—is a matter of theoretical triangulation from the data at our disposal. We should never have unalloyed confidence in the definitiveness of our data base or in the adequacy of our exploitation of it. Observation can never settle decisively just what the laws of nature are. In principle, different law systems can always yield the same observational output: as philosophers of science are wont to insist, observations *underdetermine* laws. To be sure, this worries working scientists less than philosophers, because they deploy powerful regulative principles—simplicity, economy, uniformity, homogeneity, and so on—to constrain uniqueness. But neither these principles themselves nor the uses to which they are put are unproblematic. No matter how comprehensive our data or how great our confidence in the inductions we base upon them, the potential inadequacy of our claims cannot be averted. One can never feel secure in writing *finis* on the basis of purely theoretical considerations.

We can reliably estimate the amount of gold or oil yet to be discovered, because we know a priori the earth's extent and can thus establish a proportion between what we know and what we do not. But we cannot comparably estimate the amount of knowledge yet to be discovered, because we have and can have no way of relating what we know to what we do not. At best, we can consider the proportion of available questions that we can in fact resolve; and this is an unsatisfactory procedure (see the discussion in chapter 2). The very idea of cognitive limits has a paradoxical air. It suggests that we claim knowledge about something outside knowledge. But (to hark back to Hegel), with respect to the realm of knowledge, we are not in a position to draw a line between what lies inside and what lies outside—seeing that, *ex hypothesi,* we have no cognitive access to that latter. One cannot make a survey of the relative extent of knowledge or ignorance about nature except by basing it on some picture of nature that is already in hand—that is, unless one is prepared to take at face value the deliverances of existing science. This process of judging the adequacy of our science on its own telling is the best we can do, but it remains an essentially circular and consequently inconclusive way of proceeding. The long and short of it is that there is no cognitively adequate basis for maintaining the completeness of science in a rationally satisfactory way.

To monitor the theoretical completeness of science, we accordingly need some theory-external control on the adequacy of our theorizing, some theory-external reality principle to serve as a standard of adequacy. We are thus driven to abandoning the road of pure theory and proceeding along that of the practical goals of the enterprise. This gives special importance and urgency to the pragmatic sector.

3. Pragmatic Completeness

The arbitrament of praxis—not theoretical merit but practical capability—affords the best standard of adequacy for our scientific proceedings that is available. But could we ever be in a position to claim that science has been completed on the basis of the success of its practical applications? On this basis, the perfection of science would have to manifest itself in the perfecting of control—in achieving a perfected technology. But just how are we to proceed here? Could our natural science achieve manifest perfection on the side of control over nature? Could it ever underwrite a recognizably perfected technology?

The issue of control over nature involves much more complexity than may appear on first view. For just how is this conception to be understood? Clearly, in terms of bending the course of events to our will, of attaining our ends within nature. But this involvement of *our ends* brings to light the prominence of our own contribution. For example, if we are inordinately modest in our demands (or very unimaginative), we may even achieve "complete control over nature" in the sense of being in a position to do *whatever we want* to do, but yet attain this happy condition in a way that betokens very little real capability.

One might, to be sure, involve the idea of omnipotence and construe a "perfected" technology as one that would enable us to do literally *anything*. But this approach would at once run into the old difficulties already familiar to the medieval scholastics. They were faced with the challenge: "If God is omnipotent, can he annihilate himself (contra his nature as a *necessary* being), or can he do evil deeds (contra his nature as a *perfect* being), or can he make triangles have four angles (contrary to *their* definitive nature)?" Sensibly enough, the scholastics inclined to solve these difficulties by maintaining that an omnipotent God need not be in a position to do literally anything but rather simply anything that it *is possible* for him to do. Similarly, we cannot explicate the idea of technological omnipotence in terms of a capacity to produce and result, wholly without qualification. We cannot ask for the production of a *perpetuum mobile*, for spaceships with "hyperdrive" enabling them to attain transluminar velocities, for devices that predict essentially stochastic processes such as the disintegrations of transuranic atoms, or for piston devices that enable us to set independently the values for the pressure, temperature, and volume of a body of gas. We cannot, in sum, ask of a "perfected" technology that it enable us to do anything we might take it into our heads to do, no matter how "unrealistic" this might be.

All that we can reasonably ask is that perfected technology should enable us to do anything *that it is possible for us to do*—and not just what we might think we can do but what we really and truly can do. A perfected technology would be one that enabled us to do anything that can possibly be done by creatures circumstanced as we are. But how can we deal with the pivotal conception of *can* that is at issue here? Clearly, only science—real, true, correct, perfected science—could tell us what indeed is realistically possible and what circumstances are inescapable. Whenever our knowledge falls short of this, we may well ask the impossible by way of accomplishment (for example, spaceships in hyperdrive) and thus complain of incapacity to achieve control in ways that put unfair burdens on this conception.

Power is a matter of the "effecting of things possible"—of achieving control—and it is clearly the cognitive state of the art in science that, in teaching us about the limits of the possible, is itself the agent that must shape our conception of this issue. *Every* law of nature serves to set the boundary between what is genuinely possible and what is not, between what can be done and what cannot, between which questions we can properly ask and which we cannot. We cannot satisfactorily monitor the adequacy and completeness of our science by its ability to effect "all things possible," because science alone can inform us about what is possible. As science grows and develops, it poses new issues of power and control, reformulating and reshaping those demands whose realization represents "control over nature." For science itself brings new possibilities to light. (At a suitable stage, the idea of splitting the atom will no longer seem a contradiction in terms.) To see if a given state of technology meets the condition of perfection, we must *already* have a body of perfected science in hand to tell us what is indeed possible. To validate the claim that our technology is perfected, we need to *preestablish* the completeness of our science. The idea works in such a way that claims to perfected control can rest only on perfected science.

In attempting to travel the practicalist route to cognitive completeness, we are thus trapped in a circle. Short of having supposedly perfected science in hand, we could not say what a perfected technology would be like, and thus we could not possibly monitor the perfection of science in terms of the technology that it underwrites.

Moreover, even if (per impossible) a pragmatic equilibrium between what we can and what we wish to do in science were to be realized, we could not be warrantedly confident that this condition will remain unchanged. The possibility that just around the corner things will become unstuck can never be eliminated. Even if we achieve control to all intents and purposes, we cannot be sure of not losing our grip upon it—not because of a loss of power but because of cognitive changes that produce a broadening of the imagination and a widened apprehension as to what having control involves.

Accordingly, the project of achieving practical mastery can never be perfected in a satisfactory way. The point is that control hinges on what we want, and what we want is conditioned by what we think possible, and *this* is something that hinges crucially on theory—on our beliefs about how things work in this world. Control is something deeply theory-infected. We can never safely move from apparent to real adequacy in this regard. We cannot adequately assure that seeming perfection is more than just that. We thus have no alternative but to *presume* that our knowledge (that is, our purported knowledge) is inadequate at this and indeed at any other particular stage of the game of cognitive completeness.

One important point about control must, however, be noted with care. Our preceding negative strictures all relate to attainment of perfect control—of being in a position to do everything possible. No such problems affect the issue of amelioration—of doing some things better and *improving* our control over what it was. It makes perfectly good sense to use its technological applications as standards of scientific advancement. (Indeed, we have no real alternative to using pragmatic standards at this level, because reliance on theory alone is, in the end, going to be circular.) While control does not help us with *perfection*, it is crucial for monitoring *progress*. Standards of assessment and evaluation are such that we can implement the idea of improvements (progress), but not that of completion (realized perfection). We can determine when we have managed to *enlarge* our technological mastery, but we cannot meaningfully say what it would be to *perfect* it. (Our conception of the *doable* keeps changing with changes in the cognitive state of the art, a fact that does not, of course, alter our view of what *already has been done* in the practical sphere.)

With regard to technical perfectibility, we must recognize that (1) there is no reason to expect that its realization is possible, even in principle, and (2) it is not monitorable: even if we had achieved it, we would not be able to claim success with warranted confidence. In the final analysis, then, we cannot regard the *realization* of completed science as a meaningful prospect—we cannot really say what it is that we are asking for. (To be sure, what is meaningless here is not the idea of perfected science as such but the idea of *achieving* it.) These deliberations further substantiate the idea that we must always presume our knowledge to be incomplete in the domain of natural science.

4. Predictive Completeness

The difficulties encountered in using physical control as a standard of perfection in science all also hold with respect to prediction, which, after all, is simply a mode of cognitive control.

Suppose someone asks: "Are you really still going to persist in complaints regarding the incompleteness of scientific knowledge when science can predict *everything?*" The reply is simply that science will *never* be able to predict literally everything: the very idea of predicting *everything* is simply unworkable. For then, whenever we predict something, we would have to predict also the effects of making those predictions, and then the ramification of *those* predictions, and so on *ad indefinitum*. The very most that can be asked is that science put us into a position to predict, not *everything*, but rather *anything* that we might choose to be interested in and to inquire about. Here it must be recognized that our imaginative perception of the possibilities might be much too narrow. We can only make predictions about matters that lie, at least broadly speaking, within our cognitive horizons. Newton could not have predicted findings in quantum theory any more than he could have predicted the outcome of American presidential elections. One can only make predictions about what one is cognizant of, takes note of, deems worthy of consideration. In this regard, one can be myopic either by not noting or by losing sight of significant sectors of natural phenomena.

Another important point must be made regarding this matter of unpredictability. Great care must be taken to distinguish the ontological and the epistemological dimensions, to keep the entries of these two columns apart:

unexplainable	not (yet) explained
by chance	by some cause we do not know of
spontaneous	caused in a way we cannot identify
random	lawful in ways we cannot characterize
by whim	for reasons not apparent to us

It is tempting to slide from epistemic incapacity to ontological lawfulness. But we must resist this temptation and distinguish what is inherently uncognizable from what we just don't happen to cognize. The nature of scientific change makes it inevitably problematic to slide from present to future incapacity.

Sometimes, to be sure, talk in the ontological mode is indeed warranted. The world no doubt contains situations of randomness and chance, situations in which genuinely stochastic processes are at work in ways that "engender unknowability." But these ontological claims must root in knowledge rather than ignorance. They can only be claimed appropriately in those cases in which (as in quantum theory) *we can explain inexplicability* —that is, in which we can account for the inability to predict/explain/control within the framework of a positive account of why the item at issue is actually unpredictable, unexplainable, or unsolvable.

Accordingly, these ontologically based incapacities do *not* introduce matters that lie beyond the limits of knowledge. On the contrary, positive information is the pivot point. The only viable limits to knowability are those that root in knowledge—that is, in a model of nature that entails that certain sorts of things are unknowable. It is not a matter of an incapacity to answer appropriate questions ("We 'just don't know' why that stochastic process eventuated as it did"). Rather, in the prevailing state of knowledge, these question are improper; they just do not arise.

Science itself sets the limits to predictability—insisting that some phenomena (the stochastic processes encountered in quantum physics, for example) are inherently unpredictable. This is always to some degree problematic. The most that science can reasonably be asked to do is to predict what it itself sees as in principle predictable—to answer every predictive question that it itself countenances as proper. Here we must once more recognize that any given state of science might have gotten matters quite wrong.

With regard to predictions, we are thus in the same position that obtains with regard to actually interventionist (rather than merely cognitive) control. Here, too, we can unproblematically apply the idea of improvement—of progress. It makes no sense to contemplate the *achievement* of perfection; its realization is something we could never establish by any practicable means.

5. Temporal Finality

Scientists from time to time indulge in eschatological musings and tell us that the scientific venture is approaching its end.[5] It is, of course, entirely conceivable that natural science will come to a stop and will do so not in consequence of a cessation of intelligent life but in C. S. Peirce's more interesting sense of completion of the project: of eventually reaching a condition after which even indefinitely ongoing inquiry will not—and indeed in the very nature of things cannot—produce any significant change, because inquiry has come to the end of the road. The situation would be analogous to that envisaged in the apocryphal story in vogue during the middle 1800s regarding the commissioner of the United States Patents who resigned his post because there was nothing left to invent.[6]

Such a position is in theory possible; but here, too, we can never determine that it is actual.

There is no practicable way in which the claim that science has achieved temporal finality can be validated. The question "Is the current state of science, S, final?" is one for which we can never legitimate an affirmative answer. For the prospect of future changes of S can never be precluded. One cannot plausibly move beyond "We have (in S) no good reason to think that S will ever change" to obtain "We have (in Σ) good reason to think that S will never change." To take this posture toward S is to presuppose its completeness.[7] It is not simply to take the natural and relatively unproblematic stance that that for which S

vouches is to be taken as true but to go beyond this to insist that whatever is true finds a rationalization within S. This argument accordingly embeds *finality* in *completeness*, and in doing so jumps from the frying pan into the fire. For it shifts from what is difficult to what is yet more so. To hold that if something is so at all, then S affords a good reason for it, is to take so blatantly ambitious (even megalomaniacal) a view of S that the issue of finality seems almost a harmless appendage.

Moreover, just as the appearance of erotetic and pragmatic equilibrium can be a product of narrowness and weakness, so can temporal finality. We may think that science is unchangeable simply because we have been unable to change it. But that's just not good enough. Were science ever to come to a seeming stop, we could never be sure that it had done so not because it is at "the end of the road" but because we are at the end of our tether. We can never ascertain that science has attained the w-condition of final completion, since from our point of view the possibility of further change lying just around the corner can never be ruled out finally and decisively. No matter how final a position we *appear* to have reached, the prospects of its coming unstuck cannot be precluded. As we have seen, future science is inscrutable. We can never claim with assurance that the position we espouse is immune to change under the impact of further data—that the oscillations are dying out and we are approaching a final limit. In its very nature, science "in the limit" relates to what happens in the long run, and this is something about which we *in principle* cannot gather information: any information we can actually gather inevitably pertains to the short run and not the long run. We can never achieve adequate assurance that *apparent* definitiveness is *real*. We can never consolidate the claim that science has settled into a frozen, changeless pattern. The situation in natural science is such that our knowledge of nature must ever be presumed to be incomplete.

The idea of achieving a state of recognizably completed science is totally unrealistic. Even as widely variant modes of behavior by three-dimensional objects could produce exactly the same two-dimensional shadow-projections, so very different law-systems could in principle engender exactly the same phenomena. We cannot make any definitive inferences from phenomena to the nature of the real. The prospect of perfected science is bound to elude us.

One is thus brought back to the stance of the great Idealist philosophers (Plato, Spinoza, Hegel, Bradley, Royce) that human knowledge inevitably falls short of recognizably "perfected science" (the Ideals, the Absolute) and must accordingly be looked upon as incomplete.

We have no alternative but to proceed on the assumption that the era of innovation is not over—that future science can and will prove to be different science.

As these deliberations indicate, the conditions of perfected science in point of description, explanation, prediction, and control are all unrealizable. Our information will inevitably prove inconclusive. We have no reasonable alternative to seeing our present-day science as suboptimal, regardless of the question of what date the calendar shows.

Note that the present discussion does not propound the *ontological* theses that natural science cannot be pragmatically complete, w-definitive, and so on, but the *epistemological* thesis that science cannot ever be *known to be so*. The point is not that the requirements of definitive knowledge cannot in the nature of things be satisfied but that they cannot be *implemented* (that is, be *shown* to be satisfied). The upshot is that science must always be presumed to be incomplete, not that it necessarily always is so. No doubt this is also true. It cannot, however, be demonstrated on the basis of epistemological general principles but requires the substantive considerations regarding the metaphysics of inquiry that are developed in the next chapter.

6. The Dispensability of Perfection

The cognitive situation of natural science invites description in theological terms. The ambiguity of the human condition is only too manifest here. We cannot expect ever to reach a position of definitive finality in this imperfect dispensation: we do have "knowledge" of sorts, but it is manifestly imperfect. Expelled from the garden of Eden, we are deprived of access to the God's-eye point of view. Definitive and comprehensive adequacy is denied us: we have no basis for claiming to know the truth, the whole truth, and nothing but the truth in scientific matters. We yearn for absolutes but have to settle for plausibilities; we desire what is definitively correct but have to settle for conjectures and estimates.

In this imperfect epistemic dispensation, we have to reckon with the realities of the human condition. Age disagrees with age; different states of the art involve naturally discordant conceptions and incommensurate positions. The moral of the story of the Tower of Babel applies.

The absolutes for which we yearn represent an ideal that lies beyond the range of practicable realizability. We simply have to do the best we can with the means at our disposal. To aspire to absolutes—for definitive comprehensiveness—is simply unrealistic.

It is sometimes maintained that such a fallibilist and imperfectionist view of science is unacceptable. To think of science as inevitably incomplete and to think of the definitive answers in scientific matters as perpetually unattainable is, we are told, to write science off as a meaningful project.

But in science, as in the moral life, we can operate perfectly well in the realization that perfection is unattainable. No doubt here and there some scientists nurse the secret hope of attaining some fixed and final definitive result that will stand, untouchable and changeless, through all subsequent ages, but unrealistic aspirations are surely by no means essential to the scientific enterprise as such. In science as in other domains of human endeavor, it is a matter of doing the best we can with the tools that come to hand.

For the fact that perfection is unattainable does nothing to countervail against the no less real fact that improvement is realizable—that progress is possible. The undeniable prospect of realizable progress—of overcoming genuine defect and deficiencies that we find in the work of our predecessors—affords ample impetus to scientific innovation. Scientific progress is not generated *a fronte* by the pull of an unattainable ideal; it is stimulated *a tergo* by the push of dissatisfaction with the deficiencies of achieved positions. The labors of science are not pulled forward by the mirage of (unattainable) perfection. We are pushed onward by the (perfectly realizable) wish to do better than our predecessors in the enterprise.

We can understand *progress* in two senses. On the one hand, there is O-progress, defined in terms of increasing distance from the starting point (the origin). On the other hand, there is D-progress, defined in terms of decreasing distance from the goal (the destination.) Consider the picture:

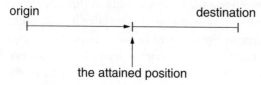

Figure 10.1 Modes of Progress

Ordinarily, the two modes of progress are entirely equivalent: we increase the distance traveled from O by exactly the same amount as we decrease the distance remaining to D. But if there is no attainable destination—if we are engaged on a journey that, for all we know, is literally endless and has no determinable destination, or only one that is "infinitely distant"—then we just cannot manage to decrease our distance from it.

Given that in natural science we are embarked on a journey that is literally endless, it is only O-progress that can be achieved, and not D-progress. We can gauge our progress only in terms of how far we have come, and not in terms of how far we have to go. Embarked on a journey that is in principle endless, we simply cannot say that we are nearing the goal.

The upshot is straightforward. The idea of *improving* our science can be implemented without difficulty, since we can clearly improve our performance as regards its goals of prediction, control, and the rest. But the idea of *perfecting* our science cannot be implemented.

7. "Perfected Science" as an Idealization that Affords a Useful Contrast Conception

Reasons of general principle block us from ever achieving a position from which we can make good the claim that the several goals of science have actually been reached. Perfection is simply not a goal or *telos* of the scientific enterprise. It is not a realizable condition of things but at best a useful contrast conception that keeps actual science in its place and helps to sensitize us to its imperfections. The validation of this idealization lies not in its future *achievability* but in its ongoing *utility* as a regulative ideal that affords a contrast to what we do actually attain—so as to highlight its salient limitations.

With respect to the moral aspirations of man's will, Kant wrote:

> Perfection [of the moral will] is a thing of which no rational being in the world of sense is at any time capable. But since it is required [of us] as practically necessary, it can be found only in an endless progress to that complete fitness; on principles of pure practical reason, it is necessary to assume such a practical progress as the real object of our will. . . . Only endless progress from lower to higher stages of moral perfection is possible to a rational but finite being.[8]

Much the same story surely holds on the side of the cognitive perfecting of man's knowledge. Here, comparable regulative demands are at work governing the practical venture of inquiry, urging us to the ever fuller realization of the potentialities of the human intellect. The discontent of reason is a noble discontent. The scientific project is a venture in self-transcendence; one of the strongest motivations of scientific work is the urge to go beyond present science—to advance the frontiers. Man's commitment to an ideal of reason in his pursuit of an unattainable systematic completeness is the epistemic counterpart of our commitment to moral ideals. It reflects a striving toward the rational that is all the more noble because it is not finally attainable. If the work of inquiring reason in the sphere of natural science were completable, this would be something utterly tragic for us.

Ideal science is not something we have in hand here and now. Nor is it something toward which we are moving along the asymptotic and approximative lines envisaged by Peirce.[9] Existing science does not and never will embody perfection. The cognitive ideals of completeness, unity, consistency, and definitive finality represent an aspiration rather than a coming reality, an idealized *telos* rather than a realizable condition of things. Perfected science lies outside history as a useful contrast case that cannot be secured in this imperfect world.

The idea of perfected science is that *focus imaginarius* whose pursuit canalizes and structures out inquiry. It represents the ultimate *telos* of inquiry, the idealized destination of a journey in which we are still and indeed are ever engaged, a grail of sorts that we can pursue but not possess. The ideal of perfection thus serves a fundamentally regulative role to mark the fact that actuality falls short of our cognitive aspirations. It marks a contrast that *regulates* how we do and must view our claims, playing a role akin to that of the functionary who reminded the Roman emperor of his mortality in reminding us that our pretensions are always vulnerable. Contemplation of this idea reminds us that the human condition is suspended between the reality of imperfect achievement and the ideal of an unattainable perfection. In abandoning this conception—in rejecting the idea of an "ideal science" that alone can properly be claimed to afford a definitive grasp of reality—we would abandon an idea that crucially regulates our view regarding the nature and status of the knowledge to which we lay claim. We would then no longer be constrained to characterize our view of things as *merely* ostensible and purported. We would be tempted to regard our picture of nature as real, authentic, and final in a manner that we at bottom realize it does not deserve.

What is being maintained here is not that completed or perfected science is a senseless idea as such but that the idea of *attaining* it is senseless. It represents a theoretically realizable state whose actual realization we can never achieve. What is unrealizable is not perfection as such but the epistemic condition of recognizing its attainment. (Even if we arrive, we can never tell that we're there!)

Does this situation not destroy the pursuit of perfection as a meaningful endeavor? Here it is useful to heed the distinction between a *goal* and an *ideal*. A goal is something that we hope and expect to achieve. An ideal is merely a wistful inkling, a "wouldn't it be nice if"—in the mode of aspiration rather than expectation. A goal motivates us in striving for its attainment; an ideal stimulates and encourages. An ideal does not provide us with a destination that we have any expectation of reaching; it is something for whose actual attainment we do not even hope. It is in *this* sense that *perfected science* is an ideal.

Here, as elsewhere, we must reckon appropriately with the standard gap between aspiration and attainment. In the practical sphere—in craftsmanship, for example, or the cultivation of our health—we may *strive* for perfection but cannot ever claim to *attain* it. And the situation in inquiry is exactly parallel with what we encounter in such other domains—ethics specifically included. The value of an ideal, even of one that is not realizable, lies not in the benefit of its attainment (obviously and *ex hypothesi*!) but in the benefits that accrue from its pursuit. The view that it is rational to pursue an aim only if we are in a position to achieve its attainment or approximation is mistaken; it can be perfectly valid (and entirely rational) if the indirect benefits of its pursuit and adoption are sufficient—if in striving after it, we realize relevant advantages to a substantial degree. An unattainable ideal can be enormously productive. Thus, the legitimation of the ideas of perfected science lies in its facilitation of the ongoing evolution of inquiry. In this domain, we arrive at the perhaps odd-seeming posture of an invocation of practical utility for the validation of an ideal.[10]

8. Science and Reality

We are now in a position to place into clearer relief one of the really big questions of philosophy: how close a relationship can we reasonably claim to exist between the answers we give to our factual questions at the level of scientific generality and precision and the reality they purport to depict?

Scientific realism is the doctrine that *science describes the real world*: that the world actually is as science takes it to be, and that its furnishings are as science envisages them to be.[11] If we want to know about the existence and the nature of heavy water or quarks, of man-eating mollusks or a luminiferous ether, we are referred to the natural sciences for the answers. On this realistic construction of scientific theorizing, the theoretical terms of natural science refer to real physical entities and describe their attributes and comportments. For example, the electron spin of atomic physics refers to a behavioral characteristic of a real, albeit unobservable, object—an electron. According to this currently fashionable theory, the declarations of science are—or will eventually become—factually true generalizations about the actual behavior of objects that exist in the world. Is this convergent realism a tenable position?

It is quite clear that it is not. There is clearly insufficient warrant for and little plausibility to the claim that the world indeed is as our science claims it to be—that we've got matters altogether right, so that *our* science is *correct* science and offers the definitive last word on the issues. We really cannot reasonably suppose that science as it now stands affords the real truth as regards its creatures of theory.

One of the clearest lessons of the history of science is that where scientific knowledge is concerned, further discovery does not just supplement but generally emends our prior information. Accordingly, we have little alternative but to take the humbling view that the incompleteness of our purported knowledge about the world entails its potential incorrectness as well. It is now a matter not simply of gaps in the structure of our knowledge, or errors of omission. There is no realistic alternative but to suppose that we face a situation of real flaws as well, of errors of commission. This aspect of the matter endows incompleteness with an import far graver than meets the eye on first view.[12]

Realism equates the paraphernalia of natural science with the domain of what actually exists. But this equation would work only if science, as it stands, has actually got it right. And this is something we are surely not inclined—and certainly not entitled—to claim. We must recognize that the deliverances of science are bound to a methodology of theoretical triangulations from the data that binds them inseparably to the state of the art of technological sophistication in data acquisition and handling.

The supposition that the theoretical commitments of our science actually describe the world is viable only if made provisionally, in the spirit of "doing the best we can now do, in the current state of the art," and giving our best estimate of the matter. The step of reification is always to be taken qualifiedly, subject to a mental reservation of presumptive revisability. We do and must recognize that we cannot blithely equate our theories with the truth. We do and must realize that the declarations of science are inherently fallible and that we can only accept them with a certain tentativeness, subject to a clear realization that they may need to be corrected or even abandoned.

These considerations must inevitably constrain and condition our attitude toward the natural mechanisms envisaged in the science of the day. We certainly do not—or should not—want to reify (hypostasize) the theoretical entities of current science, to say flatly and unqualifiedly that the contrivances of *our* present-day science correctly depict the furniture of the real world. We do not—or at any rate, given the realities of the case, should not—want to adopt categorically the ontological implications of scientific theorizing in just exactly the state-of-the-art configurations presently in hand. Scientific fallibilism precludes the claim that what we purport to be scientific knowledge is in fact real knowledge, and accordingly blocks the path to a scientific realism that maintains that the furnishings of the real world are exactly as our science states them to be. Scientific theorizing is always inconclusive.

Convergent scientific realism of the Peircean type, which pivots on the assumption of an ultimately complete and correct scientific theory (let alone those stronger versions of realism that hinge on our ability to arrive at recognizably true scientific theories), is in deep difficulty. For we have little choice but to deem science's grasps on "the real truth of things" as both tentative and imperfect.

According to one expositor, the scientific realist "maintains that if a theory has scientific merit, then we are thereby justified in concluding that ... the theoretical entities characterized by the theory really do exist."[13] This sort of position encounters insuperable difficulties. Phlogiston, caloric, and the luminiferous ether all had scientific merit in their day, but this did not establish their existence. Why, then, should things be all that different with us? Why should our scientific merit now suddenly assure actual existence? What matters for real existence is clearly (and only) the issue of truth itself, and not the issue of what is thought to be true at some particular stage of scientific history. And here problems arise. For its changeability is a fact about science that is as inductively well-established as any theory of science itself. Science is not a static system but a dynamic process.

We must accordingly maintain a clear distinction between *our conception of reality* and *reality as it really is*. Given the equation,

Our (conception of) reality = the condition of things as seen from the standpoint of "our *putative* truth" (= the truth as we see it from the vantage point of the science of the day)

we realize full well that there is little justification for holding that our present-day science indeed describes reality and depicts the world as it really is. In our heart of hearts, then, our attitude toward our science is one of guarded affirmation. We realize that there is a decisive difference between what science accomplishes and what it endeavors to accomplish.

The world *that we describe* is one thing, the world *as we describe it* is another, and they would coincide only if our descriptions were totally correct—something that we are certainly not in a position to claim. The world-as-known is a thing of our contrivance, an artifact we devise on our own terms. Even if the data uniquely determined a corresponding picture of reality and did not underdetermine the theoretical constructions we base upon them (as they always do), the fact remains that altered circumstances lead to altered bodies of data. Our recognition of the fact that the world picture of science is ever changing blocks our taking the view that it is ever *correct*.

Accordingly, we cannot say that the world *is* such that the paraphernalia of our science actually exist as such. Given the necessity of recognizing the claims of our science to be tentative and provisional, one cannot justifiably take the stance that it depicts reality. At best, one can say that it affords an estimate of it, an estimate that will presumably stand in need of eventual revision and whose creatures-of-theory may in the final analysis not be real at all. This feature of science must crucially constrain our attitude toward its deliverances. Depiction is in this regard a matter of intent rather than one of accomplishment. Correctness in the characterization of nature is achieved not by *our* science but only by *perfected* or *ideal* science—only by that (ineradicably hypothetical!) state of science in which the cognitive goals of the scientific enterprise are fully and definitively realized. There is no plausible alternative to the view that reality is depicted by *ideal* (or perfected or "complete") science, and not by the real science of the day. But, of course, it is this latter science that is the only one we've actually got—now or ever.

A viable scientific realism must therefore turn not on what our science takes the world to be like but on what ideal or perfected science takes the world to be like. The thesis that "science describes the real world" must be looked upon as a matter of intent rather than as an accomplished fact, of aspiration rather than achievement, of the ideal rather than the real state of things. Scientific realism is a viable position only with respect to that idealized science that, as we full well realize, we do not now have—regardless of the "now" at issue. We cannot justifiably be scientific realists. Or rather, ironically, we can be so only in an idealistic manner—namely, with respect to an ideal science that we can never actually claim to possess.

The posture of scientific realism—at any rate, of a duly qualified sort—is nevertheless built into the very goal structure of science. The characteristic task of science, the definitive mission of the enterprise, is to respond to our basic interest in getting the best answers we can to our questions about the world. On the traditional view of the matter, its question-resolving concern is the raison d'être of the project—to celebrate any final victories. It is thus useful to draw a clear distinction between a realism of intent and a realism of achievement. We are certainly not in a position to claim that science as we have it achieves a characterization of reality. Still, science remains unabashedly realistic in intent or aspiration. Its aim is unquestionably to answer our questions about the world correctly and to describe the world as it actually is. The orientation of science is factual and objective: it is concerned with establishing the true facts about the real world. The theories of physics purport to describe the actual operation of real entities; those Nobel prizes awarded for discovering the electron, the neutron, the pi meson, and anti-proton, the quark, and so on, were intended to recognize an enlargement of our understanding of nature, not to reward the contriving of plausible fictions or the devising of clever ways of relating observations.

The language of science is descriptively committal. At the semantic level of the content of its assertions, science makes firm claims as to how things stand in the world. A realism of intent or aspiration is built into science because of the genesis of its questions. The factually descriptive status of science is ultimately grounded in just this erotetic continuity of its issues with those of prescientific everyday life. We begin at the prescientific level of the paradigmatic realities of our prosaic everyday-life experience—the things, occurrences, and processes of our everyday world. The very reason for being of our scientific paraphernalia is to resolve our questions about this real world of our everyday experience. Given that the teleology of the scientific enterprise roots in the "real world" that provides the stage of our being and action, we are committed within its framework to take the realistic view of its mechanisms. Natural science does not address itself to some world-abstracted realm of its own. Its concern is with this familiar real world of ours in which we live and breathe and have our being—however differently science may characterize it. While science may fall short in performance, nevertheless in aspiration and endeavor it is unequivocally committed to the project of modeling the real world, for in this way alone could it realize its constituting mandate of answering our questions as to how things work in the world.

Scientific realism skates along a thin border between patent falsity and triviality. Viewed as the doctrine that science indeed describes reality, it is utterly untenable; but viewed as the doctrine that science *seeks to describe* reality, it is virtually a truism. For there is no way of sidestepping the conditional thesis: if a scientific theory regarding heavy water or electrons or quarks or whatever is correct—if it were indeed to be true—*then* its subject materials would exist in the manner the theory envisages and would have the properties the theory attributes to them: the theory, that is, would afford descriptively correct information about the world.

But this conditional relationship reflects what is, in the final analysis, less a profound fact about the nature of science than a near truism about the nature of truth as *adequation ad rem*. The fact remains that our reality—reality as we conceive it to be—goes no further than to represent our best estimate of what reality is like.

When we look to *what science declares*, to the aggregate content and substance of its declarations, we see that these declarations are realistic in intent, that they purport to describe the world as it really is. But when we look to *how science makes its declarations* and note the tentativity and provisionality with which they are offered and accepted, we recognize that this realism is of an abridged and qualified sort—that we are not prepared to claim that this is how matters actually stand in the real world. At the level of generality and precision at issue in the themes of natural science, we are not now—or ever—entitled to lay claim to the scientific truth as such but only to the scientific truth as we and our contemporaries see it. Realism prevails with respect to the language of science (that is, the asserted content of its declarations); but it should be abandoned with respect to the status of science (that is, the ultimate tenability or correctness of these assertions). What science says is descriptively committal in making claims regarding the real world, but the tone of voice in which it proffers these claims is (or should be) provisional and tentative.

Our position, then, is one not of skepticism but of realism—in two senses: (1) it is realistic about our capabilities of recognizing that here, as elsewhere, we are dealing with the efforts of an imperfect creature to do the best it can in the circumstances; and (2) it recognizes the mind-transcendent reality of a "real world" that our own best efforts in the cognitive sphere can only manage to domesticate rather imperfectly. We do, and always must, recognize that no matter how far we manage to extend the frontiers of natural science, there is more to be done. Within a setting of vast complexity, reality outruns our cognitive reach; there is more to this complex world of ours than lies—now or ever—within our ken.[14]

11 The Practical Infeasibility of Perfecting Science

Synopsis

(1) Scientific progress involves a process of technological escalation because natural science requires an ever more sophisticated technology to facilitate increasingly elaborate interactions with nature. (2) Given the constantly rising technological demands for continuing scientific progress, such advancement becomes increasingly more difficult and expensive in resource-cost terms. In a world of finite resources, this means that science must in the future progress ever more slowly—for strictly practical and ultimately economic reasons. Although natural science is theoretically limitless, its actual future development confronts obstacles and impediments of a strictly practical kind that spell its deceleration. (3) Accordingly, we will never be able to advance science as much as we would, ideally speaking, like to do. We must recognize that natural science faces not barriers (progress-stopping boundaries or absolute limits) but obstacles (progress-hampering difficulties and impediments).

1. Technological Escalation

A theoretical prospect of unending scientific progress lies before us, but its practical realization is something else again. One of the most striking and important facts about scientific research is that the ongoing resolution of significant new questions faces increasingly high demands for the generation and cognitive exploitation of data. Although the veins of gold run on, they become increasingly hard to mine.

This matter of the practical impediments to scientific progress that make themselves felt through economic pressures is generally neglected. Although our prime concern here is with theoretical rather than practical limits to science, we cannot wholly ignore this crucial issue, which demands recognition in any balanced discussion of the limits of science.

The instrumentalities of scientific inquiry can be enhanced not only on the side of theoretical resources but preeminently on the side of the technological instrumentalities of observational and experimental intervention. Scientific research operates at the technological frontier, and nature inexorably exacts an *exponentially increasing effort* with respect to the acquisition and processing of data for revealing her "secrets." This accounts for the recourse to more and more sophisticated technology for research in natural science.

Nature becomes less and less yielding to the efforts of our inquiry. As science advances, we are faced with the need to push nature harder and harder to achieve cognitively profitable interactions. That there is pay dirt deeper down in the mine avails us only if we can actually dig there. New forces, for example, may well be in the offing, if one able physicist is right:

> We are familiar, to varying degrees, with four types of force: gravity, electricity, the strong nuclear force that holds the atomic nucleus together and the weak force that brings about radioactive decay by the emission of electrons.... Yet it would indeed be astonishing if ... other types of force did not exist. Such other forces could escape our notice because they were too weak to have much distinguishable effect or because they were of such short range that, no matter whether they were weak or not, the effects specifically associated with their range were contained within the objects of the finest scale that our instruments had so far permitted us to probe.[1]

Of course, such weak forces would enter into our picture of nature only if our instrumentation was able to detect them. This need for the constant enhancement of scientifically relevant technology lies at the basis of the enormous increase in the human and material resources needed for modern experimental science.

The idea of scientific progress as the correlate of a movement through sequential stages of technological sophistication was already clearly discerned by the astute Charles Sanders Peirce around the turn of the century:

> Lamarckian evolution might, for example, take the form of perpetually modifying our opinion in the effort gradually to make that opinion represent the known facts as more and more observations came to be collected.... But this is not the way in which science mainly progresses. It advances by leaps; and the impulse for each leap is either some new observational resource, or some novel way of reasoning about the observations. Such a novel way of reasoning might, perhaps, also be considered as a new observational means, since it draws atten-

tion to relations between facts which would previously have been passed by unperceived.[2]

This circumstance has far-reaching implications for the perfectibility of science. The impetus to augment our science demands an unremitting and unending effort to enlarge the domain of effective experimental intervention. For only by operating under new and heretofore inaccessible conditions of observational or experimental systematization—attaining ever more extreme temperature, pressure, particle velocity, field strength, and so on—can we bring new grist to our scientific mill.

The perspective afforded by a model of exploration is helpful. Progress in natural science has heretofore been relatively easy because we have explored nature in our own locality: not our spatial neighborhood but our parametric neighborhood in the space of physical variables like temperature, pressure, radiation, and so on. Here, thanks to the evolutionary heritage of our sensory and cognitive apparatus, we have been able to operate with relative ease and freedom. But scientific innovation becomes more and more difficult—and expensive—as we push out farther and farther from our home base toward ever more remote frontiers.

Physicists often remark that the development of our understanding of nature moves through successive layers of theoretical sophistication.[3] But scientific progress is clearly no less dependent on continual improvements in strictly technical sophistication:

> Some of the most startling technological advances in our time are closely associated with basic research. As compared with 25 years ago, the highest vacuum readily achievable has improved more than a thousand-fold; materials can be manufactured that are 100 times purer; the submicroscopic world can be seen at 10 times higher magnification; the detection of trace impurities is hundreds of times more sensitive; the identification of molecular species (as in various forms of chromatography) is immeasurably advanced. These examples are only a small sample. . . . Fundamental research in physics is crucially dependent on advanced technology, and is becoming more so.[4]

Without an ever-developing technology, scientific progress would grind to a halt. The discoveries of today cannot be advanced with yesterday's instrumentation and techniques. To secure new observations, to test new hypotheses, and to detect new phenomena, an ever more powerful technology of inquiry is needed. Throughout the natural sciences, technological progress is a crucial requisite for cognitive progress.

Man's acquisition of knowledge about the workings of nature is clearly a matter of interaction—a transaction in which both parties, man and nature, must play a crucial role. Most writers on the limits of science operate predominantly on the side of man and see the issue as the result—primarily and in the first instance—of human failings and deficiencies (in intellect, learning power, memory, imagination, will power, etc.). It is too easily ignored that "limits" to scientific progress also reside in the physical limitations imposed upon us by the nature of the world.

We are embarked on a literally endless endeavor to improve the range of effective experimental intervention, because only by operating under new and heretofore inaccessible conditions of observational or experimental systematization—attaining extreme temperature, pressure, particle velocity, field strength, and so on—can we realize those circumstances that enable us to put our hypotheses and theories to the test. As one acute observer has rightly remarked: "Most critical experiments [in physics] planned today, if they had to be constrained within the technology of even ten years ago, would be seriously compromised."[5]

This situation points toward the idea of a "technological level," corresponding to a certain state of the art in the technology of inquiry in regard to data generation and processing. This technology of inquiry falls into relatively distinct levels or stages in sophistication—correlative with successively "later generations" of instrumentative and manipulative machinery. These levels are generally separated from one another by substantial (roughly order-of-magnitude) improvements in performance in regard to such information-providing parameters as measurement exactness, data-processing volume, detection-sensitivity, high voltages, high or low temperatures, and so on.

The key aspect of this phenomenon is that once the major findings accessible at a given data-technology level have been attained, further major progress in the problem area requires ascent to a higher level on the technological scale. Every data-technology level is subject to *discovery saturation*: the body of significant scientific findings realizable at any level is *finite* (and, indeed, not just finite but relatively small). Within a given problem area and relative to a fixed level of data technology, one thus faces a situation of eventual exhaustion, and so one obtains a localized version of the theory of progress based on the geographic-exploration model of discovery.

The exhaustion of prospects at a given level does not, of course, bring progress to a stop. For after the major findings accessible at a given data-technology level have been realized, further major findings become realizable when one ascends to the next level of sophistication in data-relevant technology. We arrive at a situation of technologic escalation, as it were. The need for new data forces one to look further and further from man's familiar home base in the parametric space of nature. Thus, while scientific progress is in principle always possible—there being no absolute or intrinsic limits to significant scientific discovery—the *realization* of this ongoing prospect demands continual enhancement in the technological state of the art of data extraction or exploitation.

Frontier research is true pioneering: what counts is not just doing but doing it *for the first time*. Aside from the initial reproduction of claimed results needed to establish the reproducibility of results, repetition in research is in general pointless. As one acute observer has remarked, one can follow the diffusion of scientific technology "from the research desk down to the schoolroom":

> The emanation electroscope was a device invented at the turn of the century to measure the rate at which a gas such as thorium loses its radioactivity. For a number of years it seems to have been used only in the research laboratory. It came into use in instructing graduate students in the mid-1930's, and in college courses by 1949. For the last few years a cheap commercial model has existed and is beginning to be introduced into high school courses. In a sense, this is a victory for good practice; but it also summarizes the sad state of scientific education to note that in the research laboratory itself the emanation electroscope has long since been removed from the desk to the attic.[6]

The enormous power, sensitivity, and complexity deployed in present-day experimental science have not been sought for their own sake but rather because the research frontier has moved on into an area where this sophistication is the indispensable requisite of ongoing progress. The choice nowadays lies between big science and no science. No doubt small is beautiful, but in natural science big becomes unavoidable: in science, as in war, one cannot fight the battles of the present with the armaments of the past.

2. Rising Costs

In any matured branch of natural science, continually greater capabilities in terms of technological capacity are required to realize further substantial results. Findings are thus "purchased" with a certain investment in scientific resources—equipment, energy, talent. And this purchase price of significant new findings constantly increases. Once all the significant findings accessible at a given state-of-the-art level of investigative technology have been realized, one must continually move on to a new, more complex (and thus expensive) level: one requires more accurate measurements, more extreme temperatures, higher voltages, more intricate combinations, and so on. The phenomenon of cost escalation is explained through a combination of the finitude of the body of substantial results realizable *within a given level* of investigative technology, together with a continual and steep increase in the resource costs of pushing from one level to the next.

One sometimes hears the cost increase in scientific work accounted for with reference to decreasing efficiency in recruitment, or to increasing personnel costs,[7] or even to boondoggling and "projecteering" by avaricious scientists. From the present perspective, such explanatory recourse to the manpower–management side of the issue seems misguided. There is no point in blaming human foibles or defective administrative arrangements for a circumstance that is built into the structure of scientific investigation (realizing, to be sure, that the facts that make bigness a necessary condition of significant progress do not establish it as sufficient). *The primary and predominant reason for the ongoing escalation in the resource cost of significant scientific discovery resides in the increasing technical difficulties in the realizing of this objective*, difficulties that are a fundamental—and ineliminable—part of an enterprise of *empirical* research in which we confront the requirement of contriving ever more far-out interactions with nature.

This cost escalation is by no means confined to the power-intensive investigations of physics. In biology, experiments that aim at detecting delicate statistical relationships, involving the control of large-scale animal populations, such as the "million mouse experiment" in genetics, indicate the lengths to which data technology will be driven in an area where massive statistics are needed. (In the social sciences, the technology of large-scale surveys affords a comparable illustration: collecting thousands of even brief interviews nowadays costs a huge sum.) The basic principles operative here are not unique to natural science. For we are concerned with an endeavor to push a technology to the limits of its capacity, and one knows from innumerable cases that there is an analogous cost increase in any situation in which technology is used to press toward any natural limit.

The preceding considerations set science clearly apart from productive enterprises of a more ordinary sort. The course of historical experience in manufacturing industries yields a picture in which (1) the industries have grown exponentially in the overall *investment* of the relevant resources of capital and labor, whereas (2) the *output* of the industries has grown at an even faster exponential rate. As a result of this combination, the ratio of investment cost per unit of output has declined exponentially due to favorable "economies of scale" throughout the manufacturing industries. However, this relationship does not hold for the science industry. The economies of mass production are unavailable in research at the scientific frontiers. Here, the existing modus operandi is always of limited utility: its potential is soon wrung dry; the frontiers keep moving onward and upward. Of course, if it were a matter of doing an experiment over and over—as in a classroom demonstration—then the unit cost could be brought down and the usual economies of scale would be obtained. The economics of mass reproduction are altogether different from those in pioneering production. With mass production, unit costs decrease as though each item made chipped off a bit of the cost of those yet to be made.[8] Here, too, an exponential relationship obtains, but one of exponential decay. However, this is by no means the situation with the development costs of *scientific* research technology. The ratio of investment per unit of output has increased exponentially in the science industry—the exact reverse of the more standard case of industrial production.[9]

In science, as in a technological arms race, one is simply never called on to keep doing what was done before. An ever more challenging task is posed by the constantly escalating demands of science for the enhanced data that can only be obtained at new levels of technological sophistication. One is always forced further up the mountain, ascending to ever higher levels of technological performance—and of cost. As science endeavors to extend its mastery over nature, it thereby comes to be involved in a technology-intensive arms race against nature, with all of the economic implications characteristic of such a process.

3. Economic Requirements Spell Economic Limitations

While we can confidently anticipate that our scientific technology will see ongoing improvement, we cannot expect it ever to attain perfection. There is no reason to think that we ever will, or indeed can, reach the end of the line. Every successive level of technical capability has its inherent limits, whose overcoming opens up yet another more sophisticated level of the technological state of the art. There is always more to be done. The accessible pressures and temperatures can in theory always be increased, the low-temperature experiments brought closer to absolute zero, the particles accelerated closer to the speed of light, and so on. Any such enhancement of practical mastery carries along (so experience teaches) new phenomena and an enhanced capability to test yet further hypotheses and discriminate between alternative theories conducive to deepening our knowledge of nature.

While there is always more to be done, the doing of it becomes increasingly difficult, and man's material resources are limited. These limits inexorably circumscribe our cognitive access to the real world. There will always be interactions with nature of such a scale (as measured in energy, pressure, temperature, particle velocities, etc.) that their realization would require the concurrent deployment of resources of so vast a scope that we can never realize them. If there are interactions to which we have no access, then there are (presumably) phenomena that we cannot discern. It would be very unreasonable to expect nature to confine the distribution of cognitively significant phenomena to those ranges that lie within our reach.

We must, accordingly, come to terms with the fact that we cannot realistically expect that our science will ever—at any given stage of its actual development—be in a position to afford us more than a very partial and incomplete control over nature. For the achievement of control over nature requires not only intellectual instrumentalities (concepts, ideas, theories, knowledge) but also, and no less importantly, the deployment of physical resources (technology and power). And the physical resources at our disposal are restricted and finite. It follows that our capacity to effect control is bound to remain imperfect and incomplete, with much in the realm of the doable always remaining undone. We shall never be able to travel down this route as we might like to go.

We must not delude ourselves into thinking that control does not matter for understanding, that *praxis* is irrelevant to *theoria*. Natural science is empirical science. Theory is intimately bound up with experimentation and observation. One of the clearest lessons of the history of science is that as we acquire more powerful means for data acquisition and processing, and thus as our information base changes, so the character of our theories, and with it our view of the world, our picture of nature, also changes. The existence of a potentially unending sequence of levels of technological sophistication entails an unending (potential) sequence of levels of theoretical sophistication, with a very different story, a different picture of nature, emerging at every level. As it gets increasingly expensive—and ultimately, in a world of limited resources, too expensive—for man to push forward the frontiers of technology, so also does it get increasingly expensive to advance the frontiers of theorizing.

Progress in modern natural science faces the challenging task of climbing ever upward from one level of technological sophistication to the next. Creative genius cannot of itself outrun the course of technological development. Nothing could more emphatically demonstrate the impotence of mere intellect unaided by technological means for the acquisition of empirical data than the fact that nowadays, in many areas of natural science, it is virtually inconceivable that major discoveries (or indeed any original work of real value and interest) should come from some quarter outside the handful of major research centers or institutes. Only they are on top of the problem at hand, and only they are privy to the new data generated by the frontier technology of research that represents a special in-house information source: particle accelerators, research reactors, radio telescopes, and so on.

Progress without new data is, of course, possible in various fields of scholarship and inquiry. The example of pure mathematics, for instance, shows that discoveries can be made in an area of inquiry that operates without empirical data. But this hardly represents a feasible prospect for natural science. It is exactly the explicit dependency on additional data—the *empirical* aspect of the discipline—that sets natural science apart not only from the *formal* sciences (logic and mathematics) but also from the *hermeneutic* ones (like the humanities), which address themselves ceaselessly to the imaginative reinterpretation and re-reinterpretation of old data from novel conceptual perspectives.

The ancient Greeks were certainly as intelligent as we are—perhaps, arguably, even more so. But given the information technology of the day, it is not just improbable but actually inconceivable that the Greek astronomers could have come up with an explanation for the red shift or the Greek physicians with an account of the bacteriological transmission of some communicable disease. The relevant types of data needed to put such phenomena within cognitive reach simply lay beyond their range. Given the instrumentalities of the times, there just was no way for the Greeks (no matter how well-endowed in brain power) to gain physical or conceptual access to the relevant phenomena. Progress in theorizing in these directions was barred, not permanently but then and there *for them*, by a technological barrier on the side of data—a barrier as absolute as the then-extant technological barriers in the way of developing the internal-combustion engine or the wireless telegraph.

The Danish historian of science A. G. Drachmann closes his excellent book *The Mechanical Technology of Greek and Roman Antiquity*[10] with the following observation: "I should prefer not to seek the cause of the failure of an invention in the social conditions till I was quite sure that it was to be found in the technical possibilities of the time." The history of *science*, as well as that of *technology*, is crucially conditioned by the limited nature of "the technical possibilities of the time."

There is no escaping the fact that—natural science being, as it is, an inescapably *empirical* enterprise—remorseless limitations are imposed upon the prospects of effective theorizing at any given stage in its development by this dependency on the available data. To say this is not to sell human ingenuity short; it is simply a matter of facing a very fundamental fact of scientific life. Progress in natural science is insuperably limited at any given time by the implicit barriers set by the available technology of data acquisition and processing. *Technological dependency sets technological limits*, first to data acquisition and then to theory projection. The achieved level of sophistication in the technological state of the art of information acquisition and processing sets definite limits to the prospects of scientific progress by restricting the range of findings that are going to be realistically accessible.

Limitations of physical capacity and capability spell cognitive limitations for empirical science. Where there are inaccessible phenomena, there must be cognitive inadequacy as well. To this extent, at any rate, the empiricists were surely right. Only the most fanatical rationalist could uphold the capacity of sheer intellect to compensate for the lack of data. The existence of unobserved phenomena means that our theoretical systematizations may well be (and presumably are) incomplete. Insofar as certain phenomena are not just undetected but in the very nature of the case inaccessible (even if only for the merely economic reasons suggested above), our theoretical knowledge of nature must be presumed imperfect. Fundamental features inherent in the structure of man's interactive inquiry into the ways of the world thus conspire to ensure the incompleteness of our knowledge.

There remains an insidiously tempting argument: the contention that a slowing in access to new phenomena under the retarding impetus of (increasingly significant) resource limitations doesn't matter all that much, because these further capabilities would in any case afford no more than icing on the cake, trivial refinements that afford minor corrections located more decimal places out. The view that underlies such a position is that further changes are smaller changes: that a juncture has been reached where the additional advances of science are merely minor details and readjustments in a basically completed picture of how nature functions.

We can here give short shrift to this mistaken view—which has already been criticized in chapter 5. There is simply no reason to think that nature conveniently assures that phenomena of difficult access are thereby insignificant in cognitive importance, so that we come near, early and easily, to exhausting the range of the cognitively significant interactions. Stanley Jevons complained a century ago:

> In the writings of some recent philosophers, especially of Auguste Comte, and in some degree John Stuart Mill, there is an erroneous and hurtful tendency to represent our knowledge as assuming an approximately complete character. At least these and many other writers fail to impress upon their readers a truth which cannot be too constantly borne in mind, namely, that the utmost successes which our scientific method can accomplish will not enable us to comprehend more than an infinitesimal fraction of what there doubtless is to comprehend.[11]

Nothing has happened in the interim to lead one to dissent from these strictures; and, for essentially economic and practical reasons, it is not likely that anything will happen in the future to alter this situation. We cannot realistically expect that our science, at *any* given stage of its actual development, will ever be in a position to afford us more than a very partial and incomplete degree of cognitive and physical control over nature. Our control is bound to be imperfect and incomplete. There is certainly no basis for an optimistic expectation that we can travel down this route as far or as fast as we might ideally like to do. In natural science, imperfect physical control is bound to mean imperfect cognitive control.

In the end, then, natural science faces not barriers (boundaries or absolute limits) but obstacles (difficulties and impediments). The technological/economic requirements for scientific progress mean that we will never be able to advance the project to our total satisfaction—will never be able to do as much as we would, ideally speaking, like to do. Given the inescapable realities of resource limitations, the prospect of a perfected or complete science is a practical impossibility. In a world of limited resources, science could certainly be *ended*—finished in the sense of being advanced as far as it is possible for creatures of our kind to push a venture of this sort—without thereby being *perfected or completed*, that is, without thereby discharging to the full the characterizing mandate of the enterprise in terms of description, explanation, prediction, and control.

If the present perspective of unremitting cost escalation in the economics of scientific progress is even partly correct, the pace of scientific advance is destined to become markedly slower in the zero-growth era that presumably lies ahead. The half-millennium that commenced around 1650 will eventually come to be regarded as among the great characteristic developmental transformations of human history, with the age of the Science Explosion seen to be as unique—and as finite—in its own historical structure as the Bronze Age and the Industrial Revolution.[12]

12. Can Computers Overcome Our Limitations?

Synopsis
(1) Some preliminary explanations are needed to clarify what problem solving involves in the present context. (2) Purely theoretical limits do not represent genuine limitations in problem solving: it is no limitation to be unable to do that which cannot possibly be done. (3) But inadequate information does become a crucial factor here. (4)–(6) And other practical limitations include real-time processing difficulties as well as matters of detail management and self-insight obstacles. (7) There are, moreover, some crucial limitations where computer determinations of computer capacity are concerned. (8) & (9) While the situation in algorithmic decision theory is somewhat different, we do here also encounter limitations by way of computer-insoluble problems. (10) Humans are situated advantageously, however, seeing that they can solve problems with computers. (11) Various difficulties with this approach must—and can—be overcome.

1. Could Computers Overcome Our Limitations?

In view of the difficulties and limitations that beset our human efforts at answering our questions in a complex world, it becomes tempting to contemplate the possibility that computers might enable us to eliminate our cognitive disabilities and to overcome those epistemic frailties of ours. And so we may wonder: are computers cognitively omnipotent? If a problem is to qualify as soluble at all, will computers always be able to solve it for us?

Of course, computers cannot bear human offspring, enter into contractual agreements, or exhibit heroism. But such processes address *practical* problems relating to the management of the affairs of human life and so do not count in the present cognitive context. Then too we must put aside *evaluative* problems of normative bearing or of matters of human affectivity and sensibility: computers cannot offer us meaningful consolation or give advice to the lovelorn. The issue presently at hand regards the capacity of computers to resolve *cognitive* problems regarding matters of empirical or formal fact. Typically, the sort of problems that will concern us here are those that characterize the sciences, in particular problems relating to the description, explanation, and prediction of the things, events, and processes that comprise the realm of physical reality. To all visible appearances computers are ideal instruments for handling the matters of cognitive complexity that arise in such contexts. The question, then, is: is there anything in the domain of cognitive problem solving that computers cannot manage to do?

The history of computation in recent times is one of a confident march from triumph to triumph. Time and again, those who have affirmed the limitedness of computers have been forced into ignominious retreat as increasingly powerful machines implementing increasingly ingenious programs have been able to achieve the supposedly unachievable. However, the question on the present agenda is not "Can computers *help* with problem solving?"—an issue that demands a resounding affirmative and needs little further discussion. There is no doubt whatever that computers can do a lot here—and very possibly more than we ourselves can. But there is an awesomely wide gap between a lot and *everything*.

First some important preliminaries. To begin with, we must, in this present context, recognize that much more is at issue with a "computer" than a mere electronic calculating machine understood in terms of its operational hardware. For one thing, software also counts. And, for another, so does data acquisition. As we here construe computers, they are electronic information-managing devices equipped with data banks and augmented with sensors as autonomous data access. Such "computers" are able not only to *process* information but also to *obtain* it. Moreover, the computers at issue here are, so we shall suppose, capable of discovering and learning and thereby able significantly to extend and elaborate their own initially programmed modus operandi. Computers in this presently operative sense are not mere calculating machines but general problem solvers along the lines of the fanciful contraptions envisioned by the aficionados of artificial intelligence. These enhanced computers are accordingly question-answering devices of a very ambitious order.

On this expanded view of the matter, we must also correspondingly enlarge our vision both of what computers can do and what reasonably can be asked of them. For it is the potential of computers as an instrumentality for universal problem solving (UPS) that concerns us here, and not merely their more limited role in the calculations of algorithmic decision theory (ADC). The computers at issue will thus be prepared to deal with factually substantive as well as merely formal (logico-mathematical) issues. This means that the questions we can ask are correspondingly diverse. For here, as elsewhere, added power brings added responsibility. The questions it is appropriate to ask thus can relate not just to matters of calculation but to the things and processes of the world.

Moreover, some preliminary discussion of the nature of problem solving is required because one has to become clear from the outset about what it is to *solve* a cognitive problem. Obviously enough, this is a matter of answering questions. Now, "to answer" a question can be construed in three ways: to offer a *possible* answer, to offer a *correct* answer, and finally to offer a *credible* answer. It is the third of these senses that will be at the center of concern here; and with good reason. For consider a problem solver that proceeds in one of the following ways: it replies "yes" to every yes/no question; or it figures out the range of possible answers and then randomizes to select one; or it proceeds by pure guesswork. Even though these so-called problem solvers may give the correct response some or much of the time, they are systematically unable to resolve our questions in the presently operative credibility-oriented sense of the term. For the obviously sensible stance calls for holding that *a cognitive problem is resolved only when an appropriate answer is convincingly provided*—that is to say, when we have a solution that we can responsibly accept and acknowledge as such. Resolving a problem is not just a matter of having an answer, and not even of having an answer that happens to be correct. The actual resolution of a problem must be credible and convincing—with the answer provided in such a way that its cogency is recognizable. In general problem solving we want not just a dictum but an *answer*—a response equipped with a contextual rationale to establish its credibility in a way accessible to duly competent recipients. To be warranted in accepting a third-party answer we must ourselves have case-specific reasons to acknowledge it as correct. A response whose appropriateness as such cannot secure rational confidence is no answer at all.[1]

With this crucial preliminary out of the way, we are ready to begin.

2. General-Principle Limits Are Not Meaningful Limitations

The question before us is: "are there *any* significant cognitive problems that computers cannot solve?" Now it must be acknowledged from the outset that certain problems are inherently unsolvable in the logical nature of things. One cannot square the circle. One cannot co-measure the incommensurable. One cannot decide the demonstrably undecidable nor prove the demonstrably unprovable. Such tasks represent absolute limitations whose accomplishment is theoretically impossible—unachievable for reasons of general principle rooted in the nature of the realities at issue.[2] And it is clear that inherently unsolvable problems cannot be solved by computers either.[3]

Other sorts of problems will not be unsolvable as such but will, nevertheless, demonstrably prove to be computationally intractable. For with respect to *purely theoretical* problems it is clear from Turingesque results in algorithmic decision theory (ADT) that there will indeed be computer insolubilia—mathematical questions to which an algorithmic respondent will give the wrong answer or be unable to give any answers at all, no matter how much time is allowed.[4] But this is a mathematical fact that obtains of necessity, and so this whole issue can be also set aside for present purposes. For in the present context of universal problem solving (UPS) the necessitarian facts of Gödel-Church-Turing incompleteness become irrelevant. Here any search for *meaningful* problem-solving limitations will have to confine its attention to problems that are in principle solvable: *demonstrably* unsolvable problems are beside the point of present concern because an inability to do what is in principle impossible hardly qualifies as a limitation, seeing that it makes no sense to ask for the demonstrably impossible.

For present purposes, then, it is limits of *capability* not limits of *feasibility* that matter. In asking about the problem-solving limits of computers we are looking to problems that *computers* cannot resolve but that other problem solvers conceivably can. The limits that will concern us here are accordingly not rooted in conceptual or logico-mathematical infeasibilities of general principle nor in absolute physical impossibilities but rather in performatory limitations imposed specifically upon computers by the world's contingent modus operandi.

And in this formulation the adverb "specifically" does real work by way of ruling out certain computer limitations as irrelevant. Things standing as they do, some problems will simply be too large given the inevitable limitations on computers in terms of memory, size, processing time, and output capacity. Suppose for the moment that we inhabit a universe which, while indeed boundless, is nevertheless finite. Then no computer could possibly solve a problem whose output requires printing more letters or numbers than there are atoms in the universe. Such problems ask computers to achieve a task that is not "substantively meaningful" in the sense that no physical agent at all—computer, organism, or whatever, could possibly achieve it. The problems that concern us here are those that are not solution-precluding on the basis of inherent mathematical or physical impossibilities. To reemphasize: our concern is with the performative limitations of computers with regard to problems that are not inherently intractable in the logical or physical nature of things.

3. Practical Limits: Inadequate Information

Often the information needed for credible problem resolution is simply unavailable. Thus no problem solver can at this point in time provide credible answers to questions like "What did Julius Caesar have for breakfast on that fatal Ides of March?" or "Who will be the first auto accident victim of the next millennium?" The information needed to answer such questions is just not available at this stage. In all problem-solving situations, the performance of computers is decisively limited by the quality of the information at their disposal. "Garbage in, garbage out," as the saying has it. But matters are in fact worse than this. Garbage can come out even where no garbage goes in.

One clear example of the practical limits of computer problem-solving arises in the context of prediction. Consider the two prediction problems set out in table 12.1. On first sight, there seems to be little difficulty in arriving at a prediction in these cases.

Table 12.1. Two Prediction Problems

Case 1	
Data	X is confronted with the choice of reading a novel by Dickens or one by Trollope. And further: X is fond of Dickens.
Problem	To predict which novel X will read.
Case 2	
Data	Z has just exactly $10.00. And further: Z promised to repay his neighbor $7.00 today. Moreover, Z is a thoroughly honest individual.
Problem	To predict what Z will do with his money.

Now suppose that we acquire some further data to enlarge our background information: pieces of information supplementary to—but nowise conflicting with or corrective of—the given premises:

Case 1: X is extremely, indeed *inordinately* fond of Trollope.
Case 2: Z also promised to repay his other neighbor the $7.00 he borrowed on the same occasion.

Note that in each case our initial information is nowise abrogated but merely enlarged by the additions in question. Nevertheless, in each case we are impelled, in the light of that supplementation, to *change* the response we were initially prepared and rationally well-advised to make. Thus when I know nothing further of next year's Fourth of July parade in Centerville U.S.A., I shall predict that its music will be provided by a marching band; but if I am additionally informed that the Loyal Sons of Old Hibernia have been asked to provide the music, then bagpipes will now come to the fore.

It must, accordingly, be recognized that the search for rationally appropriate answers to certain questions can be led astray not just by the *incorrectness* of information but by its *incompleteness* as well. The specific body of information that is actually at hand is not just important for problem resolution, it is crucial. And we can never be unalloyedly confident of problem resolutions based on incomplete information, seeing that further information can always come along to upset the apple cart. As available information expands, established problem resolutions can always become destabilized. One crucial practical limitation of computers in matters of problem solving is thus constituted by the inevitable incompleteness (to say nothing of potential incorrectness) of the information at their disposal. And here the fact that computers can only ever ingest finite—and thus incomplete—bodies of information means that their problem-resolving performance is always at risk. (Moreover, this sort of risk exists quite apart from others, such as the fact that computerized problem resolutions are always the product of many steps, each of which involves a non-zero probability of error.) If we are on a quest for certainty, computers will not help us to get there.

4. Practical Limits: Transcomputability and Real-Time Processing Difficulties

Apart from *uncomputable* (computationally irresolvable) problems there is also the range of *transcomputable* problems, problems whose computational requirements exceed the physical bounds and limits that govern the concrete realization of theoretically designed algorithmic machines.[5] Because computers are physical devices, they are subject to the laws of physics and limited by the realities of the physical universe. In particular, since a computer can process no more than a fixed number of bits per second per gram, the potential complexity of algorithms means that there is only so much that a given computer can possibly manage to do.

Then there is also the temporal aspect. To solve problems about the real world, a computer must of course be equipped with information about it. But securing and processing information is a time-consuming process and the time at issue can never be reduced to an instantaneous zero. Time-constrained problems that are enormously complex—those whose solution calls for securing and processing a vast amount of data—can exceed the reach for any computer. At some point it always becomes impossible to squeeze the needed operations into available time. There are only so many numbers that a computer can crunch in a given day. And so if the problem is a predictive one it could find itself in the awkward position that it should have started yesterday on a problem only presented to it today. Thus even under the (fact-contravening) supposition that the computer can answer *all* of our questions, it cannot, if we are impatient enough, produce those answers as promptly as we might require them. Even when given, answers may be given too late.

5. Practical Limits: Limitations of Representation in Matters of Detail Management

This situation is emblematic of a larger issue. Any computer that we humans can possibly contrive here on earth is going to be finite: its sensors will be finite, its memory (however large) will be finite, and its processing time (however fast) will be finite.[6] Moreover, computers operate in a context of finite instructions and finite inputs. Any representational model that functions by means of computers is of finite complexity in this sense. It is always a finitely characterizable system: its descriptive constitution is characterized in finitely many information-specifying steps and its operations are always ultimately presented by finitely many instructions. This array of finitudes means that a computer's modeling of the real will never capture the inherent ramifications of the natural universe of which it itself is but a minute constituent. Artifice cannot replicate the complexity of the real; reality is richer in its descriptive constitution and more efficient in its transformatory processes than human artifice can ever manage to realize. For nature itself has a complexity that is effectively endless, so that no finistic model that purports to represent nature can ever replicate the detail of reality's makeup in a fully comprehensive way, even as no architect's blueprint-plus-specifications can possibly specify *every* feature of the structure that is ultimately erected. In particular, the complications of a continuous universe cannot be captured completely via the resources of discretized computer languages. All endeavors to represent reality—computer models emphatically included—involve some element of oversimplification, and in general a great deal of it.

The fact of the matter is that reality is too complex for adequate cognitive manipulation. Cognitive friction always enters into matters of information management—our cognitive processing is never totally efficient, something is always lost in the process; cognitive entropy is always upon the scene. As far as knowledge is concerned, nature does nothing in vain and so encompasses no altogether irrelevant detail. Yet oversimplification always makes for losses, for deficiencies in cognition. Representational omissions are never totally irrelevant, so that no oversimplified descriptive model can get the full range of predictive and explanatory matters exactly right. Put figuratively, it could be said that the only "computer" that can keep pace with reality's twists and turns over time is the universe itself. It would be unreasonable to expect any computer model less complex than this totality itself to provide a fully adequate representation of it, in particular because that computer model must of course itself be incorporated *within* the universe.

6. Performative Limits of Prediction-Self-Insight Obstacles

Another important sort of practical limitation to computer problem solving arises not from the inherent intractability of questions but from their unsuitability for particular respondents. Specifically, one of the issues regarding which a computer can never function perfectly is its own predictive performance. One critical respect in which the self-insight of computers is limited arises in connection with what is known as "the Halting Problem" in algorithmic decision theory (ADC). Even if a problem is computer-solvable—in the sense that a suitable computer will demonstrably be able to find a solution by keeping at it long enough—it will in general be impossible to foretell how long a process of calculation will actually be needed. There is not—and demonstrably cannot be—a *general* procedure for foretelling with respect to a particular computer and a particular problem: "here is how long it will take to find the solution—and if the problem is not solved within this time span then it is not solvable at all." No computer can provide general insight into how long it—or any other computer, for that matter—will take to solve problems. The question "How long is long enough?" demonstrably admits of no general solution here.

And computers are—of necessity!—bound to fail even in much simpler self-predictive matters. Thus consider confronting a predictor with the problem posed by the question:

P_1: *When next you answer a question, will the answer be negative?*

This is a question that—for reasons of general principle—no predictor can ever answer satisfactorily.[7] Consider the available possibilities:

Answer given	Actually correct answer	Agreement?
YES	NO	NO
NO	YES	NO
CAN'T SAY	NO	NO

On this question, there just is no way in which a predictive computer's response could possibly agree with the actual fact of the matter. Even the seemingly plausible response "I can't say" automatically constitutes a self-falsifying answer, since in giving this answer the predictor would automatically make "No" into the response called for by the proprieties of the situation.

Here, then, we have a question that will inevitably confound any conscientious predictor and drive it into baffled perplexity. The problem poses a perfectly meaningful question to which *another* predictor could give a putatively correct answer—namely, by saying: "no—that predictor cannot answer this question at all; the question will condemn a predictor (Predictor No. 1) to baffled silence." Of course the answer "I am responding with baffled silence" is one that the initial predictor cannot cogently offer. And as to that baffled silence itself, this is something that, as such, would clearly constitute a defeat for Predictor No. 1. Still, the question that impelled Predictor No. 1 into perplexity and unavoidable failure presents no problem of principle for Predictor No. 2. And this clearly shows that there is nothing improper about that question as such. For while the question posed in P_1 will be unresolvable by a *particular* computer, and it could—in theory—be answered by *other* computers, it is not unresolvable by computers in general.

However, there are other questions that indeed are computer insolubilia for computers at large. One of them is:

P_2: *What is an example of a predictive question that no computer will ever state?*

In answering *this* question the computer would have to stake a claim of the form: "Q is an example of a predictive question that no computer will ever state." And in the very making of this claim the computer would falsify it. It is thus automatically unable to effect a satisfactory resolution. However, the question is neither meaningless nor unresolvable. A *noncomputer* problem solver could in theory answer it correctly. Its presupposition, "There is a predictive question that no computer will ever consider" is beyond doubt true. What we thus have in P_2 is an example of an in-principle solvable—and thus "meaningful"—question that, as a matter of necessity in the logical scheme of things, no problem-solving computer can ever resolve satisfactorily. The long and short of it is that every predictor—computers included—is bound to manifest versatility-incapacities with respect to its own predictive operations.[8]

However, from the angle of our present considerations, the shortcoming of problems P_1 and P_2 is that they are computer-unresolvable on the basis of theoretical general principles. It is therefore not appropriate, in the present perspective—as explained above—to count this sort of thing as a computer limitation. Are there any other, less problematic examples?

7. Performative Limits: A Deeper Look

At this point we must contemplate some fundamental realities of the situation confronting our problem-solving resources. The first of these is that no computer can ever reliably determine that all its more powerful compeers are unable to resolve a particular substantive problem (that is, one that is inherently tractable and not demonstrably unsolvable on logico-conceptual grounds).

This means that:

T_1: *No computer can reliably determine that a given substantive problem is altogether computer-irresolvable.*

This is to say that no computer can reliably determine that a particular substantive problem p is such that no computer can resolve it: $(\forall C)(\sim C \text{ res } p)$. We thus have:

$$\sim(\exists C')(\exists P)C' \det[(\forall C)(\sim C \text{ res } P)]$$

or equivalently

$$(\forall C')(\forall P)\sim C' \det[(\forall C)\sim(C \text{ res } P)].^9$$

A brief explanation is needed regarding the use of *determine* that is operative here. In the present context this is a matter of so functioning as to be able to secure rational conviction for the claim at issue. As was emphasized above, we want not just answers but *credible* answers.

Moreover, something that we are not prepared to accept from any computer is cognitive megalomania. No computer is, so we may safely suppose, ever able to achieve credibility in staking a claim to the effect that no substantive problem whatever is beyond the capacity-reach of computers. This leads to the thesis:

T_2: *No computer can reliably determine that all substantive problems whatever are computer-resolvable.*

That is to say that no computer can convincingly establish that whenever a substantive problem p is at issue, then some computer can resolve it—in other words, that for any and every substantive problem p: $(\exists C) \, C \text{ res } p$. Thus:

$$\sim(\exists C')C' \det[(\forall P)(\exists C)(C \text{ res } P)]$$

or equivalently:

$$(\forall C')\sim C' \det[\sim(\exists P)(\forall C)(\sim C \text{ res } P)]$$

Neither can a computer reliably determine that an arbitrarily given substantive problem is computer-irresolvable (T_1) nor can it reliably determine that no such problem is computer-irresolvable (T_2).

We shall not now expatiate upon the rationale of these theses. This issue of establishing their plausibility—whose pursuit would at this point unduly interrupt the flow of present deliberations—will be postponed until the appendix. All that matters at this juncture is that the principles in question merit acceptance—and do so not as a matter of abstractly mathematico-logical considerations, but owing to the world's practical realities.

The relationship between theses T_1 and T_2 comes to light more clearly when one considers their formal structure. The claims at issue are as follows:

T_1: For all C': $(\forall P) \sim C'$ det $[(\forall C)(\sim C \text{ res } P)]$

T_2: For all C': $\sim C'$ det $[(\forall P) \sim (\forall C)(\sim C \text{ res } P)]$

Now let us also adopt the following two abbreviations:
- C-un p for: $\sim C$ det P ("C is unable to determine that P")
- $X(p)$ for: $(\forall C) \sim C$ res P ("P is computer-unresolvable")

Then

$T_1 =$ For all C: $(\forall P) C$-un $X(P)$

$T_2 =$ For all C: C-un $(\forall P) \sim X(P)$

As this makes clear, both theses alike indicate a universal computer incapacity in relation to certain computer-unresolvability theses. Thus T_1 and T_2 both reflect ways in which computers encounter difficulty in obtaining a credible grip on such universal incapacity. Fixing the bounds of computer solvability is beyond the capacity of any computer.

It should be noted that in his later writings, Kurt Gödel himself took a line regarding mathematics analogous to that which the present discussion takes with respect to general problem solving. He maintained that no single particular axiomatic proof systematization will be able to achieve universality with respect to provability in general.[10] And so, even as Gödel saw algorithms as inherently incapable of doing full justice to mathematics, so the present argument has it that problem-solving computers cannot do full justice to science. Both theses alike implement the common idea that, notwithstanding the attractions and advantages of rigorous reasoning, the fact remains that in a complex world it is bound to transpire that truth is larger than rigor.

8. Contrast with Algorithmic Decision Theory

The unavailability of a universal problem solver in the setting of general problem solving has far-reaching theoretical implications. For it means that in universal problem solving (UPS) we face a situation regarding the capability of computers that is radically different from that of algorithmic decision theory (ADC).

In algorithmic decision theory we have Church's thesis:

Wherever it is possible for computation to decide an issue, this resolution can be achieved by means of effective calculation. Thus computational resolvability/

decidability (an informal conception!) can to all useful intents and purposes be equated with algorithmically effective computability (which is rigorously specifiable):[11]

(C) sol $P \leftrightarrow (\exists C)(C \text{ res } P)$.

To this thesis one can adjoin Alan Turing's great insight that there can be a "universal computer" (a "Turing machine")[12]—a device that can solve a computational problem if any calculating machine can:

(T) $(\exists C)(C \text{ res } P) \leftrightarrow T \text{ res } P$

Combining these two theses, we arrive at the result that in the sphere of algorithmic computation, solvability-at-large is tantamount to resolvability by a Turing machine:

(M) sol $P \leftrightarrow T \text{ res } P$

Here one machine can speak for the rest: if a problem is resolvable at all by algorithmic calculations, then a Turing machine can resolve it. In algorithmic decision theory (ADC), there is thus an absolute, across-the-board conception of solvability.

When we turn our perspective to universal problem solving (UPS) this monolithic situation is lost. Here the state of things is no longer Turingesque: there is not and cannot be a universal problem solver.[13] As we have seen, for any problem solver there will automatically be some correlatively unsolvable problems—problems that it cannot resolve but others can—along the lines of the aforementioned computer-embarrassing question P_1 ("Will the next answer that you give be negative?"). Once we leave the calm waters of algorithmic computation and venture into the turbulent sea of problem-solving computation in general, it becomes impracticable for any computer to survey all the possibilities. Here the overall range of computer-resolvable problems extends beyond the information horizon (the "range of vision" so to speak) of any given computer, so that no computer can make convincing claims about this range as a whole. In particular, these deliberations mean we would not—and should not—be prepared to take a computer's word for it if it stakes a claim of the format "Q is a (substantive) question that no computer whatsoever could possibly resolve."

9. A Computer Insolubilium

The time has come to turn from generalities to specifics. At this point we can confront a problem-solving computer (*any* such computer) with the challenging question:

P_3: *What is an example of a (substantive) problem that no computer whatsoever can resolve?*

There are three possibilities here:
1. The computer offers an answer of the format "P is an example of a problem that no computer whatsoever can resolve." For reasons already canvassed we would not see this as an acceptable resolution, since by T_1 our respondent cannot achieve credibility here.
2. The computer responds: "No can do: I am unable to resolve this problem: it lies outside my capability." We could—and would—accept this response and take our computer at its word. But the response of course represents no more than computer acquiescence in computer incapability.
3. The computer responds: "I reject the question as being based on an inappropriate presupposition, namely that there indeed are problems that no computer whatsoever can resolve." We ourselves would have to reject this position as inappropriate in the face of T_2. The response at issue here is one that we would simply be unable to accept at face value from a computer.

It follows from such deliberations that P_3 is itself a problem that no computer can resolve satisfactorily.

At this point, then, we have realized the principal object of the discussion: we have been able to identify a meaningful concrete problem that is computer irresolvable for reasons that are embedded—via theses T_1 and T_2—in the world's empirical realities. For—to reemphasize—our present concern is with issues of general problem solving and not algorithmic decision theory.

10. The Human Element: Can People Solve Problems that Computers Cannot?

Our discussion has not, as yet, entered the doctrinal terrain of discussions along the lines of Hubert L. Dreyfus's *What Computers Still Can't Do*.[14] For the project that is at issue there is to critique the prospects of "artificial intelligence" by identifying processes involving human intelligence and behavior that computers cannot manage satisfactorily. Dreyfus accordingly compares computer information processing with human performance in an endeavor to show that there are things that humans can do that computers cannot accomplish. However, the present discussion has to this point proceeded with a view solely to problems that computers cannot manage to resolve. Whether *humans* can or cannot resolve them is an issue that has remained out of sight.

Thus there is a big question that yet remains untouched, namely: is there any sector of this problem-solving domain where the human mind enjoys a competitive advantage over computers? Or does it transpire that wherever computers are limited, humans are always limited in similar ways?

In addressing this issue, let us be precise about the question that now faces us. It is:

P_4: *Are there problems that computers cannot solve satisfactorily but people can?*

In fact what we would ideally like to have is not just an abstract answer to P_4, but a concrete answer to:

P_5: *What is an example of a problem that computers cannot solve satisfactorily but people can?*

What we are now seeking is a computer-defeating question that has the three characteristics of (1) posing a meaningful problem, (2) being computer-unsolvable, and (3) admitting of a viable resolution by intelligent noncomputers, specifically humans.[15]

This, then, is what we are looking for; and lo and behold, *we have already found it*. All we need do is turn around and look back to P_3. After all, P_3 is—so it was argued—a problem that computers cannot resolve satisfactorily, and this consideration automatically provides us—people that we are—with the example that is being asked for. In presenting P_3 within its present context we have in fact resolved it. Moreover, P_5 is itself also a problem of just this same sort. It too is a computer-unresolvable question that people can manage to resolve.[16]

In the end, then, the ironic fact remains that the very question we are considering about cognitive problems that computers cannot solve but people can provides its own answer.[17] P_3 and P_5 appear to be eligible for membership in the category of academic questions—questions that are effectively self-resolving—a category that also includes such more prosaic members as: "What is an example of a question formulated in English?" and "What is an example of a question that asks for an example of something?" The presently operative mode of computer unsolvability thus pivots on the factor of self-reference—just as is the case with Gödelian incompleteness.

To be sure, their inability to answer the question "What is a question that no computer can possibly resolve?" is—viewed in a suitably formidable perspective—a token of the power of computers rather than of their limitedness. After all, we see the person who maintains "I can't think of something I can't accomplish" not as unimaginative but as a megalomaniac—and one who uses "we" instead of "I" as only slightly less so. Nevertheless, in the present case this pretention to strength marks a point of weakness.

The key issue is whether computers might be defeated by questions that other problem solvers, such as humans, could overcome. The preceding deliberations indicate that there indeed are such limitations. For the ramifications of self-reference are such that one computer could satisfactorily answer certain questions regarding the limitation of the general capacity of computers to solve questions. But humans can in fact resolve such questions because, with them, no self-reference is involved.

11. Potential Difficulties

The time has now come for facing up to some possible objections and difficulties.

An objection that deserves to be given short shrift runs as follows: "but the sort of computer insolubilium represented by P_5 is really not the kind of thing I was expecting when contemplating the title of the chapter." Expectations do not really count for much in this sort of situation. After all, nobody faults Gödel for not having initially coming up with the sort of examples that people might have expected regarding the incompleteness of formalized arithmetic—some insoluble diophantine problem in number theory.[18]

But of course other objections remain.

For example, do those instanced problems really lie outside the domain of trustworthy computer operation? Could computers not simply follow in the wake of our own reasoning here and instance P_3 and P_5 as self-resolving? Not really. For in view of the considerations adduced in relation to T_1-T_2 above, a computer cannot convincingly monitor the range of computer-tractable problems. Thus the responses of a computer in this sort of issue simply could not secure rational conviction.

But what if a computer were to establish its reliability regarding such supposedly computer-irresolvable questions indirectly? What about a reliable black box? Could a computer not acquire reliability simply by compiling a good track record?

Well . . . yes and no. A black box can indeed establish credibility by compiling a good track record of correct responses. But it can do so only when this track record is issue-homogeneous with the matter at hand: when those correct responses relate to questions of the same type as the one that is at issue. Credibility is not transferable from botany to mathematics or from physics to theology. The only productively meaningful track record would have to be one compiled *in a reference class of similar cases*. Now just how does type-homogeneity function with respect to our problem? What is the "type of problem" that is at issue here? The answer is straightforward: it is *questions that for reasons of principle qualify as being computer-intractable*. But how could a computer establish a good track record here? Only by systematically providing the responses we can reasonably deem to be correct on wholly independent grounds. The situation that arises here would thus be analogous to that of a black box that systematically forecasts tomorrow's headlines correctly. This sort of thing is indeed imaginable—it is a logically feasible possibility (and thereby, no doubt, an actuality in the realm of science fiction). But we would be so circumstanced as to deem that black box's performance as miraculous. And we do not—cannot—accept this as a *practical* possibility for the real world. It is a fanciful hypothesis that we would reject out of hand until such time as actual circumstances confronted us with its realization—a prospect we would dismiss as utterly unrealistic. It represents a bridge that we would not even think about crossing until we actually got there—simply because we have a virtually ineradicable conviction that actually getting there is something that just will not happen.

"But surely the details of this discussion are not so complex that a computer capable of defeating grand masters at chess could not handle them as well." There thus still remains a subtle and deep difficulty. One recent author formulated the problem as follows:

> In a way, those who argue for the existence of tasks performable by people and not performable by computers are forced into a position of never-ending retreat. If they can specify just what their task involves, then they [must] admit the possibility of programming it on some machine. . . . . [And] even if they construct proofs that a certain class of machines cannot perform certain tasks, they are vulnerable to the possibility of essentially new classes of machines being described or built.[19]

After all, when people can solve a certain problem successfully, then they can surely "teach" this solution to a computer by simply adding the solution to its information store. And thereafter the computer can also solve the problem by replicating the solution—or if need be by simply looking it up.

Well, so be it. It is certainly possible for a computer to maintain a solution registry. Every time some human solves a problem somewhere somehow, it is duly entered into this register. And then the computer can determine the person-solvability of a given problem by the simple device of going and "looking it up." But this tactic represents a hollow victory. First of all, it would give the computer access only to person-resolved problems and not to person-resolvable ones. But, more seriously yet, if a computer needs *this* sort of input for answering a question, then we could hardly characterize the problem at issue as computer-solvable in any but a Pickwickian sense.

At this point the issue of the scoring system becomes crucial. We now confront what is perhaps the most delicate and critical part of the inquiry. For now we have to face and resolve the conceptual question of how the attribution of credit works in matters of problem solving.

Clearly if *all* of the inferential steps essential to establishing a problem solution as such were computer-performed, and *all* of the essential data inputs were computer-provided, then computers will have to be credited with that problem solution. But what of the mixed cases where some essential contributions were made on both sides—some by computers and some by people? Here the answer is that *credit for mixed solutions lies automatically with people.* For if a computer "solves" the problem in a way that is overtly and essentially dependent on people-provided aid, then its putative "solution" can no longer count as authentically computer-provided. A problem that is "not solvable by persons alone but yet solvable when persons are aided by computers" is still to be classed as person-solvable, while a problem that is "not solvable by computers alone but yet solvable when computers are aided by persons" is not to be classed as computer-solvable. (Or at any rate is not so until we reach the science-fiction level of self-produced, self-programmed, independently evolving computers that manage to reverse the master-servant relationship.) For as matters stand, the scoring system used in these matters is simply not "fair." The seemingly table-turning question "Is there a problem that people cannot solve but computers can?" automatically requires a negative response once one comes to realize that *people can and do solve problems with computers.*[20] The conception of "computer-provided solutions" works in such a way that here computers must not only do the job but actually *the whole job.* And on this basis the difficulty posed by that subtle objection can be dismissed.

The crucial point is that while people use computers for problem solving, the converse simply does not hold: the prospect that computers solve problems by using people as investigative instruments is unrealistic—barring a technologico-cultural revolution that sees the emergence of functionally autonomous computers, operative on their own and able to press people into their service.

Does such a principle of credit allocation automatically render people superior to computers? Not necessarily. Quite possibly the things that computers cannot accomplish in the way of problem solving are things people cannot accomplish either—be it with or without their means. The salient point is surely that much of what we would ideally like to do, computers cannot do either. They can indeed diminish but cannot eliminate our limitations in solving the cognitive problems that arise in dealing with a complex world, that is, in effect, a realm where *every* problem-solving resource faces some ultimately insuperable obstacles.[21]

Appendix to Chapter 12: On the Plausibility of T_1 and T_2

The task of this appendix is to set out the plausibility considerations that establish the case for accepting the pivotal theses T_1 and T_2.

A helpful starting point for these deliberations is provided by the recognition that the inherently progressive nature of pure and applied science ensures the prospect of continual improvements in the development of ever more capable instrumentalities for general problem solving. No matter how well we are able to do at any given state-of-the-art stage in this domain the prospect of further improvements always lies open. Further capabilities in point of information access and/or processing capacity can always be added to any realizable computer, no matter how powerful it may be. And this suffices to substantiate the realization that there is no inherent limit here: for every particular problem solver that is actually realized there is (potentially) some other whose performative capability is greater.[22]

Now it lies in the nature of things that in cognitive matters, an agent possessed of a lesser range of capabilities will always underperform one possessed of a greater range. More powerful problem solvers can solve more problems. A chess player who can look four moves ahead will virtually always win out over one who can manage only three moves. A crossword puzzler who can manage words of four syllables will surpass one who can manage only three. A mathematician who has mastered the calculus will outperform one whose competency is limited to arithmetic. This sort of thing holds for general problem solving as well.

One must also come to terms with the realization that no problem solver can ever reliably determine that all its more powerful compeers are unable to resolve a particular substantive problem (that is, one that is inherently tractable and not demonstrably unsolvable on logico-conceptual grounds). The plausibility argument for this is straightforward and roots in the limited capacity of a feebler intelligence to gain adequate insight into the operation of a stronger. After all, one of the most fundamental facts of epistemology is that a lesser capacity can never manage to comprehend fully the operations of a greater. The untrained mind cannot grasp the ways of the expert. Try as one will, one can never adequately translate Shakespeare into pidgin English. Similarly, no problem solver can determine the limits of what its more powerful compeers can accomplish. None can reliably resolve questions of computer solvability in general: none can reliably survey the entire range. The enhanced performance of a more capable intellectual performer will always seem mysterious and almost magical to a less capable compeer.

John von Neumann conjectured that computational complication is *reproductively* degenerative in the sense that computing machines can only *produce* others less complicated than themselves. The present thesis is that with regard to universal problem solving (UPS), computational complication is *epistemically* degenerative in that computing machines can only reliably *comprehend* others less complicated than themselves (where one computer "comprehends" another when it is in a position to tell just what this other can and cannot do).

Considerations along these lines substantiate that no computer in the field of general problem solving can obtain a secure overview of the performance of its compeers at large. This means that:

Thesis T_1: *No computer can reliably determine that a given substantive problem is altogether computer-irresolvable.*

Furthermore, it is clear that the question "But even if computers had no limits, how could a computer possibly manage to determine that this is the case?" plants an ineradicable shadow of doubt in our minds. For even if it were the case that computers had no problem-solving limits, and even if a computer could—as is virtually inconceivable—manage to determine that this is so, the fact would nevertheless remain that the computer could not really manage to secure our conviction with respect to this claim. Trusting though we might be, we would not—and could not reasonably—be *that* trusting. No computer could achieve credibility in staking so hyperbolic a claim. (We have the sort of situation that reminds one of the old Roman dictum: "I would not believe it even were it told to me by Cato.")

The upshot is that no computer problem solver is in a position to settle the question of limits that affect its compeers across the board. We thus have it that:

Thesis T_2: *No computer can reliably determine that all substantive problems whatever are computer-resolvable.*

But of course this thesis—like its predecessor—holds only in a domain that, like universal problem solving (UPS), is totally open-ended. It does not hold for algorithmic decision theory (ADC) where one single problem solver (the Turing machine) can speak for all the rest.

The following dialectical stratagem also deserves notice. Suppose someone were minded to contest the acceptance of theses T_1 and T_2 and proposed to reject one of them. This very stance of theirs would constrain them to concede the other. For T_2 *follows from the denial of* T_1 (and correspondingly T_1 follows from the denial of T_2). In other words, there is no prospect of denying both these theses; at least one of them *must* be accepted.

The proof here runs as follows. Let $X(p)$ as usual represent $(\forall C) \sim (C \text{ res } P)$. Then we have:

T_1: For all C: $(\forall P) \sim (C \text{ det } X(P))$

T_2: For all C: $\sim (C \text{ det } (\forall P) \sim X(P))$

We thus have:

$\sim T_1$ = For some C: $(\exists P)(C \text{ det } X(P))$

$\sim T_2$ = For some C: $C \text{ det } (\forall P) \sim X(P)$ or equivalently $C \text{ det} \sim (\exists P) X(P)$

Now the computers at issue are supposed to be truth-determinative, so that we stand committed to the idealization that C det P entails P. On this basis, $\sim T_1$ yields $(\exists P)X(P)$. And furthermore $\sim T_2$ yields $\sim(\exists P)X(P)$. Since these are logically incompatible, we have it that $\sim T_1$ and $\sim T_2$ are incompatible so that $\sim T_1$ entails T_2 (and consequently $\sim T_2$ entails T_1).

It is a clearly useful part of the plausibility argumentation for T_1-T_2 to recognize that accepting at least one of them is inescapable.

13 Extraterrestrial Science

(Could Aliens Overcome Our Limitations?)

Synopsis
(1) Might another, astronomically remote, alien civilization surpass our human science and become "scientifically more advanced" than we are? (2) A negative answer is indicated: the "science" of alien beings is bound to be very different from ours. (3) A critique is made of the uniformitarian thesis that since there is only one world that all intelligent beings share, there is bound to be only one uniform science. Even though there might well be enormously many intelligent civilizations in space, the probability that any have our scientific posture is negligibly small. (4) It only makes sense to speak of being "more advanced" or "more backward" than another when the parties are engaged in a common journey. This is hardly likely to be so in the present case. (5) Cognition is an evolutionary product that is bound to attune its practitioners to the local peculiarities of their particular ecological niche in the world order. (6) We must assume that their intellectual journey of reasoned inquiry will take them in an altogether different direction. Our science is limited by the very fact of being our science. It is thus far-fetched to suppose that an alien civilization might be scientifically more advanced than we are.

1. Could Science in Another Setting Overcome the Limitations of Our Human Science?

The preceding chapters have argued that natural science—*our* science as we humans cultivate it here on earth—is limited and imperfect and is bound to remain so. It thus becomes tempting to wonder whether an astronomically remote civilization might be scientifically more advanced than we are. Is it not plausible to suppose that an alien civilization might overcome the limitations of our science and manage to surpass us in the furtherance of this enterprise?

On first thought, the question seems very clear-cut, for, as one recent discussion put it: "any serious speculations concerning the capabilities of intelligent biological life and automata must take into account technical societies that may be millions or even billions of years more advanced than our own."[1] However, this seemingly straightforward matter is actually one of great complexity. This complexity relates not only to the actual or possible facts of the matter but also—and crucially—to somewhat abstruse questions about the very idea that is at issue here.

To begin with, there is the question of just what it means for there to be another science-possessing civilization. Note that this is a question that *we* are putting—a question posed in terms of the applicability of *our* term, *science*. It pivots on the issue of whether *we* would be prepared to consider certain of *their* activities—once we understood them—as engaged in scientific inquiry, and whether *we* would be prepared to recognize the product of their activities as constituting a (state of a branch of) science. At the very least, this requires that we be prepared to recognize what those aliens are doing as a matter of forming beliefs (theories) about how things work in the world and to acknowledge that they are involved in testing these beliefs observationally or experimentally and applying them in practical (technological) contexts. We must, to begin with, be prepared to accept them as (nonhuman) *persons*, duly equipped with intellect and will, and we must then enter upon a complex series of claims with respect to their beliefs and their purposes. And even their being intelligent persons need not carry them as far as science. Montaigne tells us that nature's noblemen, the recently discovered inhabitants of Brazil, "passoyent leur vie en une admirable simplicité et ignorance, sans lettres, sans loy, sans roy, sans religion quelconque"("manage very well without the usual accoutrements of what Europeans think of as civilized").[2] And of course they managed quite well without science too.

A scientific civilization is not merely one that possesses intelligence and social organization but one that puts this intelligence and organization to work in a very particular way. This opens up a rather subtle issue of priority in regard to process versus product. Is what counts for a civilization's "having a science" primarily a matter of the substantive *content* of their doctrines (their belief structures and theory complexes), or is it primarily a matter of the *aims and purposes* with which their doctrines are formed?

The matter of content turns on the issue of how similar their scientific beliefs are to ours, which is clearly something on which we would be ill-advised to put much emphasis. After all, the speculations of the nature theorists of pre-

Socratic Greece, our ultimate ancestors in the scientific enterprise, bear little resemblance to our present-day sciences, nor does the content of contemporary physics bear all that much resemblance to that of Newton's day. We would do better to give prime emphasis to matters of process and purpose.

Accordingly, the matter of these aliens "having a science" is to be regarded as turning not on the extent to which their *findings* resemble ours but on the extent to which their *project* resembles ours: we must decide whether we are engaged in the same sort of rational inquiry in terms of the sorts of issues being addressed and the ways in which they are going about addressing them. The issue is at bottom not one of the *substantive similarity* of their science to ours but one of the *functional equivalency* of their projects to the scientific enterprise as we know it. Only if they are pursuing such goals as description, explanation, prediction, and control of nature will they be doing *science*.

2. The Potential Diversity of "Science"

To what extent would the *functional equivalent* of natural science built up by the inquiring intelligences of an astronomically remote civilization be bound to resemble our science? In reflecting on this question and its ramifications, one soon comes to realize that there is an enormous potential for diversity.

To begin with, the *machinery of formulation* used in expressing their science might be altogether different. Specifically, their mathematics might be very unlike ours. Their dealings with quantity might be entirely anumerical—purely comparative, for example, rather than quantitative. Especially if their environment is not amply endowed with solid objects or stable structures congenial to measurement—if, for example, they were jellyfishlike creatures swimming about in a soupy sea—their "geometry" could be something rather strange, largely topological, say, and geared to flexible structures rather than fixed sizes or shapes. Digital thinking might be undeveloped, while certain sorts of analogue reasoning might be highly refined. Or, again, an alien civilization might, like the ancient Greeks, have "Euclidean" geometry without analysis. In any case, given that the mathematical mechanisms at their disposal could be very different from ours, it is clear that their description of nature in mathematical terms could also be very different. (Not necessarily truer or falser, but just different.)

Secondly, the *orientation* of the science of an alien civilization might be very different. All their efforts might conceivably be devoted to the social sciences—to developing highly sophisticated analogues of psychology and sociology, for example. In particular, if the intelligent aliens were a diffuse assemblage of units comprising wholes in ways that allow of overlap,[3] then the role of social concepts might become so paramount for them that nature would throughout be viewed in fundamentally social categories, with those aggregates we think of as physical structures contemplated by them in social terms.

Then, too, their natural science might deploy mechanisms very different from ours. Communicating by some sort of "telepathy" based upon variable odors or otherwise "exotic" signals, they might devise a complex theory of emphatic thought-wave transmittal through an ideaferous ether. Again, the aliens might scan nature very differently. Electromagnetic phenomena might lie altogether outside the ken of alien life forms; if their environment does not afford them lodestone and electrical storms, the occasion to develop electromagnetic theory might never arise. The course of scientific development tends to flow in the channel of practical interests. A society of porpoises might lack crystallography but develop a very sophisticated hydrodynamics; one comprised of molelike creatures might never dream of developing optics or astronomy. One's language and thought processes are bound to be closely geared to the world as one experiences it. As is illustrated by the difficulties we ourselves experience in bringing the language of everyday experience to bear on subatomic phenomena, our concepts are ill-attuned to facets of nature different in scale or structure from our own. We can hardly expect a science that reflects such parochial preoccupations to be a universal fixture.

The interests of creatures shaped under the remorseless pressure of evolutionary adaptations to very different—and endlessly variable—environmental conditions might well be oriented in directions very different from anything that is familiar to us.

Laws are detectable regularities in nature. But detection will of course vary drastically with the mode of observations—that is, with the sort of resources that different creatures have at their disposal to do their detecting. Everything depends on how nature pushes back on our senses and their instrumental extensions. Even if we detect everything we can, we will not have got hold of everything available to others. And the converse is equally true. The laws that we (or anybody else) manage to formulate will depend crucially on one's place within nature—on how one is connected into its wiring diagram, so to speak.

A comparison of the sciences of different civilizations here on earth suggests that it is not an outlandish hypothesis to suppose that the very *topics* of alien science might differ dramatically from those of ours. In our own case, for example, the fact that we live on the surface of the earth (unlike whales), the fact that we have eyes (unlike worms) and thus can *see* the heavens, the fact that we are so situated that the seasonal positions of heavenly bodies are intricately connected with agriculture—all these facts are clearly connected with the development of astronomy. The fact that those distant creatures would experience nature in ways very different from ourselves means that they can be expected to raise very different sorts of questions. Indeed, the mode of emplacement within nature of alien inquirers might be so different as to focus their attention on entirely different aspects of constituents of the cosmos. If the world is sufficiently complex and multifaceted, they might concentrate upon aspects of their environment that mean nothing to us, with the result that their natural science is oriented in directions very different from ours.[4]

Moreover, the *conceptualization* of an alien science might be very different, for we must reckon with the theoretical possibility that a remote civilization might operate with a drastically different system of concepts in its cognitive dealings with nature. Different cultures and different intellectual traditions, to say nothing of different sorts of creatures, are bound to describe and explain their experiences—their world as they conceive it—in terms of concepts and categories of understanding substantially unlike ours. They would diverge radically with respect to what the Germans call their *Denkmittel*—the conceptual instruments they employ in thought about the facts (or purported facts) of the world. They could, accordingly, be said to operate with different conceptual schemes, with different conceptual tools used to make sense of experience—to characterize, describe, and explain the items that figure in the world as they view it. The taxonomic and explanatory mechanisms by means of which their cognitive business is transacted might differ so radically from ours that intellectual contact with them would be difficult or impossible.

Epistemologists have often said things to the effect that people whose experience of the world is substantially different from our own are bound to conceive of it in very different terms. Sociologists, anthropologists, and linguists talk in much the same terms, and philosophers of science have recently also come to say the same sorts of things. According to Thomas Kuhn, for example, scientists who work within different scientific traditions—and thus operate with different descriptive and explanatory "paradigms"—actually "live in different worlds."[5]

It is (or should be) clear that there is no simple, unique, ideally adequate concept-framework for describing the world. The botanist, horticulturist, landscape gardener, farmer, and painter will operate from diverse cognitive points of view to describe one selfsame vegetable garden. It is merely mythology to think that the phenomena of nature can lend themselves to only one correct style of descriptive and explanatory conceptualization. There is surely no ideal scientific language that has a privileged status for the characterization of reality. Different sorts of creatures are bound to make use of different conceptual schemes for the representation of their experience. To insist on the ultimate uniqueness of science is to succumb to the myth of the God's-eye view. Different cognitive perspectives are possible, no one of them more adequate or more correct than any other independent of the aims and purposes of their users.

Supporting considerations for this position have been advanced from very different points of view. One example is a *Gedankenexperiment* suggested by Georg Simmel in the last century, which envisaged an entirely different sort of cognitive being: intelligent and actively inquiring creatures (animals, say, or beings from outer space) whose experiential modes are quite different from our own.[6] Their senses respond rather differently to physical influences: they are relatively insensitive, say, to heat and light but substantially sensitized to various electromagnetic phenomena. Such intelligent creatures, Simmel held, could plausibly be supposed to operate within a largely different framework of empirical concepts and categories; the events and objects of the world of their experience might be very different from those of our own: their phenomenological predicates, for example, might have altogether variant descriptive domains. In a similar vein, William James wrote: "Were we lobsters, or bees, it might be that our organization would have led to our using quite different modes from these [actual ones] of apprehending our experiences. It *might* be too (we cannot dogmatically deny this) that such categories unimaginable by us to-day, would have proved on the whole as serviceable for handling our experiences mentally as those we actually use."[7]

The science of members of a different civilization would inevitably be closely tied to the particular pattern of their interaction with nature as funneled through the particular course of their evolutionary adjustment to their specific environment. The "forms of sensibility" of radically different beings (to invoke Kant's useful idea) are likely to be radically diverse from ours. The direct chemical analysis of environmental materials might prove highly useful, and bioanalytic techniques akin to our senses to taste and smell could be very highly developed, providing them with environmentally oriented experiences of a very different sort.

The constitution of alien inquirers—physical, biological, and social—thus emerges as crucial for science. It would be bound to condition the agenda of questions and the instrumentalities for their resolution—to fix what is seen as interesting, important, relevant, and significant. Because it determines what is seen as an appropriate question and what is judged as an admissible solution, the cognitive posture of the inquirers must be expected to play a crucial role in shaping the course of scientific inquiry itself.

To clarify this idea of a conceptually different science, it helps to cast the issue in temporal rather than spatial terms. The descriptive characterization of *alien* science is a project rather akin in its difficulty to that of describing our own *future* science. It is a key fact of life that progress in science is a process of *ideational* innovation that always places certain developments outside the intellectual horizons of earlier workers. The very concepts we think in terms of become available only in the course of scientific discovery itself. Like the science of the remote future, the science of remote aliens must be presumed to be such that we really could not achieve intellectual access to it on the basis of our own position in the cognitive scheme of things. Just as the technology of a more advanced civilization would be bound to strike us as magic, so its science would be bound to strike us as incomprehensible gibberish—until we had learned it from the ground up. They might (just barely) be able to *teach* it to us, but they could not *explain* it to us by transposing it into our terms.

The most characteristic and significant difference between one conceptual scheme and another arises when the one scheme is committed to something the other does not envisage at all—something that lies outside the conceptual range of the other. A typical case is that of the stance of Cicero's thought-world with regard to questions of quantum electrodynamics. The Romans of classical antiquity did not hold *different* views on these issues; they held no view at all about them. This whole set of relevant considerations remained outside their conceptual repertoire. The diversified history of *our* terrestrial science gives one some minuscule inkling of the vast range of possibilities along these lines.

The science of different civilizations may well, like Galenic and Pasteurian medicine, simply *change the subject* in key respects so as to no longer talk about the same things but deal with materials (e.g., humors and bacteria, respectively) of which the other takes not cognizance at all. The difference in regard to a conceptual scheme between modern and Galenic medicine is not that the modern physician has a different theory of the operation of the four humors from his Galenic counterpart but that modern medicine has *abandoned* the four humors, and not that the Galenic physician says different things about bacteria and viruses but that he says *nothing* about them.

As long as the fundamental categories for the characterization of thought—the modes of spatiality and temporality, of structural description, functional connection, and explanatory rationalization—are not seen as necessary features of intelligence as such, but as evolved cognitive adaptations to particular contingently constituted modes of emplacement in and interaction with nature, there will be no reason to expect uniformity. Sociologists of knowledge tell us that even for us humans here on earth, our Western science is but one of many competing ways of conceptualizing the world's processes. And when one turns outward toward space at large, the prospects of diversity become virtually endless. It is a highly problematic contention even that beings constituted as we are and located in an environment such as ours must inevitably describe and explain natural phenomena in our terms. And with differently constituted beings, the basis of differentiation is amplified enormously. Our minds are the information-processing mechanisms of an organism interacting with a particular environment via certain particular senses (natural endowments, hardware) and certain culturally evolved methods (cultural endowments, software). With different sorts of beings, these resources would differ profoundly—and so would the cognitive products that would flow from their employment.

The more one reflects on the matter, the more firmly one is led to the realization that our particular human conception of the issues of science is something parochial, because we are physically, perceptually, and cognitively limited and conditioned by our specific situation within nature. Given intelligent beings with a physical and cognitive nature profoundly different from ours, one simply cannot assert with confidence what the natural science of such creatures would be like.

3. The One-World, One-Science Argument

One writer on extraterrestrial intelligence poses the question "What can we talk about with our remote friends?" and answers with the remark: "We have a lot in common. We have mathematics in common, and physics, and astronomy."[8] Another maintains that "we may fail to enjoy their music, understand their poetry, or approve their ideals; but we can talk about matters of practical and scientific concern."[9] But is it all that simple? With respect to his hypothetical Planetarians, the ingenious Christiaan Huygens wrote, three centuries ago:

> Well, but allowing these Planetarians some sort of reason, must it needs be the same with ours? Why truly I think 'tis, and must be so; whether we consider it as applied to Justice and Morality, or exercised in the Principles and Foundations of Science.... For the aim and design of the Creator is every where the preservation and safety of his Creatures. Now when such a reason as we are masters of, is necessary for the preservation of Life, and promoting of Society (a thing that they be not without, as we shall show) would it not be strange that the Planetarians should have such a perverse sort of Reason given them, as would necessarily destroy and confound what it was designed to maintain and defend? But allowing Morality and Passions with those Gentlemen to be somewhat different from ours, ... yet still there would be no doubt, but that in the search after Truth, in judging of the consequences of things, in reasoning, particularly in that form which belongs to Magnitude or Quantity about which their Geometry (if they have such a thing) is employed, there would be no doubt I say, but that their Reason here must be exactly the same, and go the same way to work with ours, and that what's true in one part will hold true over the whole Universe; so that all the difference must lie in the degree of Knowledge, which will be proportional to the Genius and Capacity of the inhabitants.[10]

With a timely shift from a theological to a natural-selectionist rationale, this analysis is close to the sort of thing one hears advanced today.

It is tempting to reason: "Since there is only one nature, only one science of nature is possible." Yet, on closer scrutiny, this reasoning becomes highly problematic. Above all, it fails to reckon with the fact that while there indeed is only one world, nevertheless very different *thought worlds* can be at issue in the elaborations of a science.

It is surely naive to think that because one single object is in question, its description must issue in one uniform result. This view ignores the crucial impact of the describer's intellectual orientation. Minds with different concerns and interests and with different experiential backgrounds can deal with the selfsame items in ways that yield wholly disjoint and disparate results because different features of the thing are being addressed. The *things* are the same, but their significance is altogether different.

Perhaps it seems plausible to argue thus: "Common problems constrain common solutions. Intelligent alien civilizations have in common with us the problem of cognitive accommodation to a shared world. Natural science as we know it is *our* solution of this problem. Therefore, it is likely to be *theirs* as well." But this tempting argument founders on its second premise. The problem situation confronted by extraterrestrials is *not* common with ours. Their situation must be presumed to be substantially different exactly because they live in a significantly different environment and come equipped with significantly different resources—physical and intellectual alike. The common problems, common solutions line does not work: to presuppose a common problem is already to beg the question.

Science is always the result of *inquiry* into nature, and this is inevitably a matter of a *transaction* or *interaction* in which nature is but one party and the inquiring beings another. We must expect alien beings to question nature in ways very different from our own. On the basis of an *interactionist* model, there is no reason to think that the sciences of different civilizations will exhibit anything more than the roughest sorts of family resemblance.

Our alien colleagues scan nature for regularities, using (at any rate, to begin with) the sensors provided to them by their evolutionary heritage. They note, record, and transmit those regularities that they find to be useful or interesting, and then develop their inquiries by theoretical triangulation from this basis. Now, this is clearly going to make for a course of development that closely gears their science to their particular situation—their biological endowment ("their sensors"), their cultural heritage ("what is pragmatically useful"). Where these key parameters differ, we must expect that the course of scientific development will differ as well.

Admittedly, there is only one universe, and its laws and materials are, as far as we can tell, the same everywhere. We share this common universe with all life forms. However radically we differ in other respects (in particular, those relating to environment, to natural endowments, and to style or civilization), we have a common background of cosmic evolution and a common heritage of natural laws. And so, if intelligent aliens investigate nature at all, they will investigate the same nature we ourselves do. All this can be agreed. But the fact remains that the corpus of scientific information—ours or anyone's—is an ideational construction. And the sameness of the object of contemplation does nothing to guarantee the sameness of ideas about it. It is all too familiar a fact that even where only human observers are at issue, very different constructions are often placed upon the "same" occurrences. As is clearly shown by the rival interpretations of different psychological schools—to say nothing of the court testimony of rival "experts"—there need be little uniformity in the conceptions held about one selfsame object from different perspectives of consideration. The fact that all intelligent beings inhabit the same world does not countervail the no less momentous fact that we inhabit very different ecological niches within it, engendering very different sorts of modus operandi.

The universality and intersubjectivity of our science, its repeatability and investigator-independence, still leave matters at the level of *human* science. As C. S. Peirce was wont to insist, the aim of scientific inquiry is to allay *our* doubts—to resolve the sorts of questions we ourselves deem worth posing. Different sorts of beings might well ask very different sorts of questions.

No one who has observed how very differently the declarations of a single text (the Bible, say, or the dialogues of Plato) have been interpreted and understood over the centuries—even by people of a common culture heritage—can be hopeful that the study of a common object by different civilizations must lead to a uniform result. Yet, such textual analogies are oversimple and misleading, because the scientific study of nature is not a matter of decoding a preexisting text. There just is not one fixed basic text—the changeless book of nature writ large—that different civilizations can decipher in different degrees. Like other books, it is to some extent a mirror: what looks out depends on who looks in.

The development of a science—a specific codification of the laws of nature—always requires as input some inquirer-supplied element of determination. The result of such an interaction depends crucially on the contribution from both sides—from nature and from the intelligences that interact with it. A kind of chemistry is at work in which nature provides only one input and the inquirers themselves provide another—one that can massively and dramatically affect the outcome in such a way that we cannot disentangle the respective contributions of nature and the inquirer. Things cannot of themselves dictate the significance that an active intelligence can attach to them. Human organisms are essentially similar, but there is not much similarity between the medicine of the ancient Hindus and that of the ancient Greeks.

After all, throughout the earlier stages of man's intellectual history, different human civilizations developed their natural sciences in substantially different ways. The shift to an extraterrestrial setting is bound to amplify this diversity. The science of an alien civilization may be far more remote from ours than the language of our cousin the dolphin is remote from our language. We must face, however reluctantly, the fact that on a cosmic scale the hard physical sciences have something of the same cultural relativity that one encounters with the softer social sciences on a terrestrial basis.

There is no categorical assurance that intelligent creatures will *think* alike in a common world, any more than that they will *act* alike—that is, there is no reason why *cognitive* adaptation should be any more uniform than *behavioral* adaptation. Thought, after all, is simply a kind of action; and as the action of a creature reflects its biological heritage, so does its mode of thought.

These considerations point to a clear lesson. Different civilizations composed of different sorts of creatures must be expected to create diverse "sciences." Although inhabiting the same physical universe with us, and subject to the same sorts of fundamental regularities, they must be expected to create as cognitive artifacts different depictions of nature, reflecting their different modes of emplacement within it.

Each inquiring civilization must be expected to produce its own, perhaps ever-changing, cognitive products—all more or less adequate in their own ways but with little if any actual overlap in conceptual content.

Natural science—broadly construed as inquiry into the ways of nature—is something that is in principle endlessly plastic. Its development will trace out a historical course closely geared to the specific capacities, interests, environment, and opportunities of the creatures that develop it. We are deeply mistaken if we think of it as a process that must follow a route generally parallel to ours and issue in a roughly comparable product. It would be grossly unimaginative to think that either the journey or the destination must be the same—or even substantially similar.

Factors such as capacities, requirements, interests, and course of development are bound to affect the shape and substance of the science and technology of any particular space-time region. Unless we narrow our intellectual horizons in a parochially anthropomorphic way, we must be prepared to recognize the great likelihood that the "science" and "technology" of a remote civilization would be something *very* different from science and technology as we know it. Our human sort of natural science may well be *sui generis,* adjusted to and coordinated with a being of our physical constitution, inserted into the orbit of the world's processes and history in our sort of way. It seems that in science, as in other areas of human endeavor, we are prisoners of the thought world that our biological and social and intellectual heritage affords us.

4. Comparability and Judgments of Relative Advancement

We have insisted from the outset that *science* must be understood in terms not of the substantive content of assertions but of a rather generic sort of *functional* equivalency of enterprises. The question that concerns us at present, however, is not whether a remote civilization has a "science" of some sort but whether it is *scientifically more advanced* than ours. Now, if *advancement* is to be at issue, and the question is to be one of *relative* sophistication, then an actual substantive comparison with our science must be provided for. If the natural science of an alien civilization is to represent an *advance* over ours, we must clearly construe it as *our sort* of science in a much more substantive way.

Even a superficial scrutiny of the terrestrial situation suffices to show that the development of natural science is certainly not inevitable for cognitive beings but is closely linked to their internal constitution and their cultural orientation. And given the immense diversity to be expected among the various modes of science and technology, the number of extraterrestrial civilizations possessing a science and technology that are duly consonant and contiguous with ours—and, in particular, oriented toward the mathematical laws of the electromagnetic spectrum—must be judged to be very small.

It must be recognized that sciences can vary: (1) in their formal mechanisms of *formulation*—their "mathematics," (2) in their *conceptualization,* that is, in the kinds of explanatory and descriptive concepts they bring to bear, and (3) in their *orientation* toward the manifold pressures of nature, reflecting the varying "interest"-directions of their developers. While science as such is clearly not anthropocentric, science *as we have it*—the only science that we ourselves know—is a specifically human artifact that must be expected to reflect to a significant degree the particular characteristics of its makers. Consequently, the prospect that an alien science-possessing civilization has a science that we would acknowledge (if sufficiently informed) as representing the same general inquiry as that in which we ourselves are engaged seems extremely implausible. The possibility that *their* science and technology are sufficiently similar in orientation and character to be substantively proximate to *ours* must be viewed as extremely remote.

Just such comparability with our sorts of science is, however, the indispensable precondition for judgments of relative advancement of backwardness vis-à-vis ourselves. The idea of their being scientifically "more advanced" is predicated on the uniformity of the enterprises: doing better and more effectively the kinds of things that *we* want science and technology to do. Any talk of advancement and progress is predicated on the sameness of the direction of movement: only if others are traveling along the same route as we are can they be said to be ahead of or behind us. The issue of relative advancement is linked inseparably to the idea of doing the same sort of thing better or more fully. One can say that a child's expository writing is more primitive than an adult's, or that a novice's performance at arithmetic or piano playing is less developed than that of an expert. But we can scarcely say that Chinese cookery is more or less advanced than Roman, or Greek pottery than Renaissance glassware. The salient point for present purposes is simply that where the enterprises are sufficiently diverse, the ideas of relative advancement and progress are inapplicable for lack of a sine qua non condition. Of course, even doing the same sort of thing at a very remote level of generality may not suffice for the comparability judgments of doing it better or worse that are needed to underwrite judgments of relative progressiveness. The condition at issue is merely necessary and not sufficient. Butterflies and sparrows both fly, but it makes little sense to say that one does so better than the other—that these are "more advanced" fliers than those.

Our discussion has occasionally compared spatial and temporal remoteness. But in this context of advancement or progressiveness, there arises a crucial disanalogy between *alien* science and *future* science. Future science is always comparable with ours, being the successor-state of a successor-state of a successor-state. And advantageousness is transmitted along this chain of succession. If a state of science were not pragmatically dominant over one already in place, it would (presumably) not establish itself as a successor. The very conditions of historic change in our terrestrial science assure its progressiveness in terms of prediction and control—and thus presumptively in terms of explanatory adequacy as well. No such in-principle assurance of relative comparability is available in the case of an alien "science."

The matter of the aliens' entitlement to claim scientific superiority is not as simple as it may seem at first sight. To begin with, scientific superiority does not automatically emerge from their capacity to make many splendidly successful predictions. For this could be the result of precognition, say, or empathetic attunement to nature. What is wanted is not just a matter of correct but of cognitively underwritten and thus science-guided predictions.

In the same way, technological wonders do not necessarily reflect *scientific* accomplishments. After all, bees can do all sorts of things that we would like to do but cannot. The technology in question must clearly be the product of intelligent contrivance rather than evolutionary trial and error. To betoken advancement, their performatory wonders must actually issue from (superior) theoretical knowledge—that is, from superior science.

Nor would the matter of scientific superiority be settled by the consideration that an extraterrestrial species might be more "intelligent" than we are in having a greater capacity for the efficient and comprehensive monitoring and processing of information. After all, whales or porpoises, with their larger brains, may (for all we know) have to manipulate substantially larger quantities of sheer data than we do, to maintain effective adaptation within a highly changeable environment. What clearly counts for scientific knowledge is not the *quantity* of intelligence in sheer volumetric terms but its *quality* in substantive, issue-oriented terms. Information manipulation does not assure scientific development. Libraries of information (or misinformation) can be generated about trivia—or dedicated to matters very different from science as we know it.

It is perhaps too tempting for humans to reckon cognitive superiority by the law of the jungle, judging as superior those who do or would come out on top in outright conflict. But surely the Mongols were not possessors of a civilization scientifically superior to that of the Near Eastern cultures they decimated in medieval times. We earthlings might easily be eliminated by not very knowledgeable creatures who are able to emit at will—perhaps à la H. G. Wells's *War of the Worlds*—a naturally secreted chemical agent capable of killing us off. *Physical* superiority may have nothing to do with *scientific* superiority.

Moreover, we might find it difficult or impossible to appraise the extent to which those aliens achieve cognitively based control over their environment. For we might lack the conceptual apparatus needed to recognize what they consider control. If their purposes are anomalous by our standards, or if their ways of pursuing them are inscrutable from our standpoint, they may well achieve control in ways we cannot appreciate. Contrarily, when they seem to us to achieve control, this may be due not to scientific intelligence but to fortuitous circumstances—or even to the fact that we simply misread their purposes and credit them with inappropriate accomplishments.

The main point, then, is that if they are to effect an *advance* on our science, they must both (1) do roughly our sort of thing in roughly our sort of way, and (2) do it significantly better. The "science" of another civilization will be "more advanced" than our own only if they have developed science (*our* sort of science—science as we know it) further than we have ourselves. And this is implausible. Even assuming that "they" develop a "science" at all—that is, a *functional equivalent* of our science—it seems unduly parochial to suppose that they are at work constructing *our* sort of science in a *substantive* sense.

Diverse organisms have diverse requirements; diverse requirements engender diverse technologies; diverse technologies make for diverse styles of science. If a civilization of intelligent aliens develops a science at all, it seems reasonable to expect that they will develop it in another direction altogether and produce something that we, if we could come to understand it at all, would regard as simply disjoint in content—although presumably not in intent—from science as we ourselves cultivate it.

There just is no unique itinerary of scientific or technological development that different civilizations travel in common with mere differences in speed or in staying power (notwithstanding the penchant of astrophysicists for the neat plotting of numerical "degrees of development" against time in the evolution of planetary civilizations).[11] In cognitive and even in "scientific" evolution, we are not dealing with a single railway line but with a complex network leading to mutually remote destinations. Even as cosmic evolution involves a spatial red shift that carries different star systems ever farther from each other, so cognitive evolution may well involve an intellectual red shift that carries different civilizations into thought worlds ever more remote from each other.

The literature created by extraterrestrial-intelligence enthusiasts is pervaded by the haunting worry: "Where is everybody?" "Why haven't we heard from them?" Are they simply too distant—or perhaps too cautious[12] or too detached?[13] Our present discussion offers yet another line of response: they are simply too busy doing their own thing. Radio communication is ours, theirs is something very, *very* different. If alien civilizations inhabit alien thought worlds, then this lack of intellectual communion might well explain the lack of physical communication.

If "being there" in scientific terms means having *our* sort of scientifically guided technology and our sort of technologically channeled science, then it does not seem all that far-fetched to suppose that *we might be there alone*—even in a universe teeming with other intelligent civilizations. The prospect that an alien civilization is going about the job of doing *our* science—a science that reflects the sorts of interests and involvement that *we* have in nature—better than we do it ourselves must accordingly be judged as extremely far-fetched.

There are, no doubt, various ways in which the prospect of alien civilizations might cause us legitimate concern. But their being scientifically more advanced that we are is not one of them. It only makes sense to speak of being "more advanced" or "more backward" when the parties are engaged in a common journey. This is hardly likely to be so in the present case.

5. First Principles

The rationale for our analysis emerges from the data of table 13.1. What matters here is not the numerical detail but the general structure. For these figures, interestingly, embody the familiar situation that as one moves through successive stages of increasing complexity, one encounters a greater scope for diversity: further layers of system complexity provide for an ever-widening spectrum of possible states and conditions. (The more fundamental the system, the narrower its correlative range of alternatives; the more complex, the wider.) If each unit ("letter," "cell," "atom") can be configured in ten ways, then each ordered group of ten such units ("words," "organs," "molecules") can be configured in

10^{10} ways, and each complex of ten such groups ("sentences," "organisms," "objects") in $(10^{10})^{10}$ — 10^{100} ways. Thus even if only a small fraction of what is realizable in theory is realizable in nature, any increase in organizational complexity will nevertheless be accompanied by an enormous amplification of possibles.

To be sure, the specific particulars of the various computations that comprise the quantitative thread of the discussion cannot be given much credence, but their general tendency nevertheless conveys an important lesson. People frequently seem inclined to reason as follows:

There are, after all, an immense number of planetary objects running about in the heavens. Proper humility requires us to recognize that there is nothing all that special about the Earth. If it can evolve life and intelligence and civilization and science, then so can other planets. And given that there are so many other runners in the race, we must assume that—even though we cannot see them in the cosmic darkness—some of them have got ahead of us in the race.

As one recent writer formulates this familiar argument: "Since man's existence on the earth occupies but an instant in cosmic time, surely intelligent life has progressed far beyond our level on some of these 100,000,000 (habitable) planets [in our galaxy]."[14] Such reasoning overlooks the critical probabilistic dimension. Admittedly, cosmic locales are very numerous. But when probabilities get to be very small, they will offset this fact. (No matter how massive N is, there is always that diminutive $1/N$ able to countervail it.) Even though there

Table 13.1. Conditions for the Development of Science.

Planets of sufficient size for potential habitation	10^{22}
Fraction thereof with the right astrophysics for a temperature location	10^{-1}
Fraction thereof with the right chemistry for life support	10^{-1}
Fraction thereof with the right biochemistry for the actual emergence of life	10^{-2}
Fraction thereof with the right biology and psychology for the evolution of intelligence	10^{-4}
Fraction thereof with the right sociology for developing a culture with a duly constructed "technology" and "science"	10^{-7}
Fraction thereof with the right epistemology for developing science as we know it	10^{-7}
Product of all these fractions	10^{-22}

are an immense number of solar systems, and thus a staggering number of sizable planets (some 10^{22} by current estimates), nevertheless, a very substantial number of conditions must be met for "science" (as we understand it) to arise. The astrophysical, physical, chemical, biological, psychological, sociological, and epistemological parameters must all be in proper adjustment. There must be habitability, and life, and intelligence, and culture, and technology, and a technologically geared mode of inquiry, and an appropriate subject-matter orientation of this intellectual product, and so on. By this reckoning, the number of civilizations that possess a technologized science as we comprehend it is clearly not going to be very substantial: it might, in fact, be strikingly close to one.

The developmental path from intelligence to science is strewn with substantial obstacles. Matters must be propitious not just as regards the physics, chemistry, biochemistry, evolutionary biology, and cognitive psychology of the situation; the sociological requisites for the evolution of science as a cultural artifact must also be met. Economic conditions, social organization, and cultural orientation must all be properly adjusted before the move from intelligence to science can be accomplished. For scientific inquiry to evolve and flourish, there must, in the first place, be cultural institutions whose development requires specific economic conditions and favorable social organizations. And terrestrial experience suggests that such conditions for the social evolution of a developed culture are by no means always present where intelligence is. At this point, we would do well to recall that of the hundreds of human civilizations evolved here on earth, only one, the Mediterranean/European, managed to inaugurate natural science as we understand it. The successful transition from intelligence to science is certainly not a sure thing.

A great many turnings must go all right en route for science of a quality comparable to ours to develop. Each step along the way is one of finite (and often small) probability. And to reach the final destination, all these probabilities must be multiplied together, yielding a quantity that will be very small indeed. Even if there were only twelve turning points along this developmental route, each involving a chance of successful eventuation that is, on average, no worse than one in one hundred, the chance of an overall success would be diminutively small, corresponding to an aggregate success-probability of merely 10^{-24}.

George G. Simpson has rightly stressed the many chance twists and turns that lie along the evolutionary road, insisting that

> the fossil record shows very clearly that there is no central line leading steadily, in a goal-directed way, from a protozoan to man. Instead there has been continual and extremely intricate branching, and whatever course we follow through

the branches there are repeated changes both in the rate and in the direction of evolution. Man is the end of one ultimate twig.... Even slight changes in earlier parts of the history would have profound cumulative effects on all descendant organisms through the succeeding millions of generations.... The existing species would surely have been different if the start had been different, and if any stage of the histories of organisms and their environments had been different. Thus the existence of our present species depends on a very precise sequence of causative events through some two billion years or more. Man cannot be an exception to this rule. If the causal chain had been different, *homo sapiens* would not exist.[15]

The workings of evolution—be it of life or intelligence or culture or technology or science—are always the product of a great number of individually unlikely events. Any evolutionary process involves putting to nature a sequence of questions whose successive resolution produces a series reminiscent of the game Twenty Questions, sweeping over a possibility-spectrum of awesomely large proportions. The result eventually reached lies along a route that traces out one particular contingent path within a possibility-space that encompasses an ever-divergent fanning out of alternatives as each step opens up yet further eventuations. An evolutionary process is a very iffy proposition—a complex labyrinth in which a great many twists and turns in the road must be taken aright for matters to end up as they do.

If things had not turned out suitably at each stage, we would not be here to tell the tale. The many contingencies on the long route of cosmic, galactic, solar-system, biochemical, biological, social, cultural, and cognitive evolution have all turned out right; the innumerable obstacles have all been surmounted. In retrospect, it all looks easy and inevitable. The innumerable (unrealized) possibilities of variation along the way are easily kept out of sight and out of mind. The Whig interpretation of history beckons comfortably. It is so easy, so tempting, to say that a planet on which there is life will of course evolve a species with the technical capacity for interstellar communications.[16] It is tempting, but it is also nonsense. There are simply too many critical turnings along the road of cosmic and biological evolution. The fact is that many junctures along the way are such that, had things gone only a little differently, we would not be here at all.[17]

The ancient Greek atomists' theory of possibility affords an interesting lesson in this connection. Adopting a Euclideanly infinitistic view of space, they held to a theory of innumerable worlds:

> There are innumerable worlds, which differ in size. In some worlds there is no sun and moon, in others they are larger than in our world, and others more numerous. The intervals between the worlds are unequal; in some parts there

are more worlds, in others fewer; some are increasing, some at their height, some decreasing; in some parts they are arising, in others failing. They are destroyed by collision one with another. There are some worlds devoid of living creatures or plants or any moisture.[18]

On this basis, the atomists taught that every (suitably general) possibility is realized in fact someplace or other. Confronting the question of "Why do dogs not have horns; just why is the theoretical possibility that dogs be horned not actually realized?" the atomist replied that it indeed is realized, but just elsewhere—*in another region of space*. Somewhere within infinite space, there is another world just like ours in every respect save one: that its dogs have horns. That dogs lack horns is simply a parochial idiosyncrasy of the particular local world in which we interlocutors happen to find ourselves. Reality accommodates all possibilities of world alternative to this through spatial distribution: as the atomists saw it, *all* alternative possibilities are in fact actualized in the various subworlds embraced within one spatially infinite superworld.

This theory of virtually open-ended possibilities was shut off by the closed cosmos of the Aristotelian world picture, which dominated European cosmological thought for almost two millennia. The breakup of the Aristotelian model in the Renaissance and its replacement by the Newtonian model is one of the great turning points of the intellectual tradition of the West—elegantly portrayed in Alexandre Koyré's splendidly entitled book, *From the Closed World to the Infinite Universe*.[19] One may recall Giordano Bruno's near-demonic delight with the explosion of the closed Aristotelian world into one opening into an infinite universe spread throughout endless space. Others were not delighted but appalled: John Donne spoke of "all coherence lost," and Pascal was frightened by "the eternal silence of infinite spaces," of which he spoke so movingly in the *Pensées*. But no one doubted that the onset of the Newtonian world picture represented a cataclysmic event in the development of Western thought.

Strangely enough, the refinitization of the universe effected by Einstein's general relativity produced scarcely a ripple in philosophical or theological circles, despite the immense stir caused by other aspects of the Einsteinian revolution. (Einsteinian space-time is, after all, even more radically finitistic than the Aristotelian world picture, which left open, at any rate, the prospect of an infinite future with respect to time.)

To be sure, it might perhaps seem that the finitude in question is not terribly significant because the distances and times involved in modern cosmology are so enormous. But this view is rather naive. The difference between the finite and the infinite is as big as differences can get to be. And it represents a difference that is—in this present context—of the most far-reaching significance. For it means that we have no alternative to supposing that a highly improbable set of eventuations is not going to be realized in very many places, and that something sufficiently improbable may well not be realized at all. The decisive *philosophical* importance of cosmic finitude lies in the fact that in a finite universe only a finite range of alternatives can be realized. A finite universe must "make up its mind" about its contents in a far more radical way than an infinite one. And this is particularly manifest in the context of low-probability possibilities. In a finite world, unlike an infinite one, we cannot avoid supposing that a prospect that is sufficiently unlikely is simply not going to be realized at all, that in piling improbability on improbability we eventually outrun the reach of the actual. It is therefore quite conceivable that our science represents a solution of the problem of cognitive accommodation that is terrestrially locale-specific.

6. The Implausibility of Being Outdistanced

Our science is bound to be limited in crucial respects by the very fact of its being *our* science. A tiny creature living its brief life span within a maple leaf could never recognize that such leaves are deciduous—themselves part of a cyclic process. The processes of this world of ours (even unto its utter disappearance) could make no cognitive impact upon a being in whose body our entire universe is but a single atom. No doubt the laws of our world are (part of) the laws of its world as well, but this circumstance is wholly without practical effect. Where causal processes do not move across the boundaries between worlds— where the levels of relevantly operative law are so remote that nothing happening at the one level makes any substantial impact on the other—there can be little if any overlap in science. Science is limited to the confines of discernibility: as Kant maintained, the limits of our experience set limits to our science.

A deep question arises: is the mission of intelligence in the cosmos uniform or diversified? Two fundamentally opposed philosophical views are possible with respect to cognitive evolution in its cosmic perspective. The one is a *monism* that sees the universal mission of intelligence in terms of a certain shared destination, the attainment of a common cosmic "position of reason as such." The other is a *pluralism* that sees each intelligent cosmic civilization as foregoing its own characteristic cognitive destiny and takes it as the mission of intelligence as such to span a wide spectrum of alternatives and to realize a vastly diversified variety of possibilities, with each thought form achieving its own peculiar destiny in separation from all the rest. The conflict between these doctrines must in the final analysis be settled triangulation from the empirical data. This said, it must be recognized that the whole tendency of these present deliberations is toward the pluralistic side. It seems altogether plausible to see cognition as an evolutionary product that is bound to attune its practitioners to the characteristic peculiarities of their particular niche in the world order.

There is, no doubt, a certain charm to the idea of companionship. It would be comforting to reflect that, however estranged from them we are in other ways, those alien minds share *science* with us at any rate and are our fellow travelers on a common journey of inquiry. Our yearning for companionship and contact runs deep. It might be pleasant to think of ourselves not only as colleagues but as junior collaborators whom other, wiser minds might be able to help along the way. Even as many in sixteenth-century Europe looked to those strange pure men of the Indies (East or West) who might serve as moral exemplars for sinful European man, so we are tempted to look to alien inquirers who surpass us in scientific wisdom and might assist us in overcoming our cognitive deficiencies. The idea is appealing, but it is also, alas, very unrealistic.

In the late 1600s, Christiaan Huygens wrote:

> For 'tis a very ridiculous opinion that the common people have got among them, that it is impossible a rational Soul should dwell in any other shape than ours.... This can proceed from nothing but the Weakness, Ignorance, and rejoice of Men; as well as the humane Figure being the handsomest and most excellent of all others, when indeed it's nothing but a being accustomed to that figure that makes me think so, and a conceit ... that no shape or colour can be so good as our own.[20]

People's tendency to place all rational minds into a physical structure akin to their own is paralleled by a tendency to emplace all rational knowledge into a cognitive structure akin to their own. Roland Pucetti even thinks that the fundamental legal and social concepts of extraterrestrial societies must be designed on our lines.[21]

Life on other worlds might be very different from the life we know. It could well be based on a multivalent element other than carbon and be geared to a medium other than water—perhaps even one that is solid or gaseous rather than liquid. In his splendid book entitled *The Immense Journey*, Loren Eiseley wrote: "Life, even cellular life, may exist out yonder in the dark. But high or low in nature, it will not wear the shape of man. That shape is the evolutionary product of a strange, long wandering through the attics of the forest roof, and so great are the chances of failure, that nothing precisely and identically human is likely ever to come that way again."[22]

What holds for the material configuration of the human shape would seem no less applicable to the cognitive configuration of human thought. It is plausible to think that alien creatures will solve the problems of *intellectual* adjustment to their environment in ways as radically different from ours as those by which they solve the problems of physical adjustment. The physics of an alien civilization need resemble ours no more than does their physical therapy. We must be every bit as leery of *cognitive* anthropomorphism as of *structural* anthropomorphism. (Fred Hoyle's science-fiction story entitled *The Black Cloud* is thought-provoking in this regard.[23] The cloud tells a scientist what it knows about the world. The result is schizophrenia and untimely death for the scientist: the cloud's information is divergent and compelling.)

With respect to biological evolution it seems perfectly sensible to reason as follows: "what can we say about the forms of life evolving on these other worlds? . . . [I]t is clear that subsequent evolution by natural selection would lead to an immense variety of organisms; compared to them, all organisms on Earth, from molds to men, are very close relations."[24] The same situation will surely obtain with respect to cognitive evolution. The "sciences" produced by different civilizations here on earth—the ancient Chinese, Indians, and Greeks for example—unquestionably exhibit an infinitely greater similarity than obtains between our present-day science and anything devised by astronomically remote alien civilizations. The idea of a comparison in terms of "advance" or "backwardness" is highly implausible. The prospect that some astronomically remote civilization is "scientifically more advanced" than ourselves—that somebody else is doing "our sort of science" *better* than we ourselves—requires in the first instance that they be doing our sort of science at all. And this deeply anthropomorphic supposition is extremely dubious. (It should be stressed, however, that this consideration that "our sort of natural science" may well be unique is not so much a celebration of our intelligence as a recognition of our peculiarity.)

The chance that an alien civilization might develop *our* sort of science as it stands in the year 2000 is remote, to put it mildly. Aliens might well surpass us in many ways—in power, in longevity, in intelligence, in ferocity, and so on. But to worry (or hope) that they might surpass us in science as we understand it is to orient one's concern in an unprofitable direction.

Nevertheless, these deliberations have a profound and far-reaching bearing on the theme of the limits of science. The ultimate reason why we cannot expect alien intelligences to be at work doing our sort of science is that the possible sorts of science are almost endlessly diverse. Sciences—understood as such in the functional-equivalency terms laid down for the present discussion—are bound to vary with the cognitive instruments that are available through the physical constitution and mental mindset of their developers and with the cognitive focus of interest of their cultural perspective and conceptual framework. In view of the world's inherent complexity we must think of our sort of science as simply one alternative among others: our whole cognitive project is simply the intellectual product of one particular sort of cognitive life form.

Empiricist philosophers of science worry whether the set of actual observations might fail to do justice to "the facts." They fret, for example, lest the characterization of reality be undermined by actual observations. And they seek to remove these worries by turning from the *actual* to the totality of *all possible* observations. But if *possible* here means *possible for us humans*—with our particular sort of observational technology—then it seems clear that we may well still be left with something inherently fragmentary and incomplete, something that reflects the characteristic limitations of a limited creature.

It follows that even if we could somehow manage (*per impossible*) to bring *our* sort of science to perfection, there is no reason to think that it would yield the definitive truth about nature as any other inquiring being (let alone God!) would ultimately conceptualize it. We could not really be in a position to claim that we have completed science per se. There is here yet another whole dimension of "limitation" to which our natural science is subject—inherent in the consideration that (as best we can judge the matter) it represents a characteristically *human* enterprise.

Immanuel Kant's insight holds: there is good reason to think that natural science as we know it is not something universally valid for all rational intelligences as such but a man-made creation correlative with our specifically human intelligence. We have little alternative to supposing that our science is limited precisely by its being *our* science. The potential plurality of modes of judgment means that there is no definitive way of "knowing the world." It entails the modesty of a cognitive Copernicanism that recognizes that our position is no more central in the cognitive than in the spatial order of things.

At the same time it encourages confidence in the idea that nobody else in the cosmos is doing *science* as we understand it better than ourselves. On this basis we can be reasonably confident that our limitations—the inherent limitations that confront *us* in the pursuit of *our* scientific project—are not something that alien beings in the cosmos would manage to overcome.

Appendix: References for Chapter 13

Note: This listing is confined to those relevant materials that I have found particularly interesting or useful. It does not aspire to comprehensiveness. A much fuller bibliography is given in MacGowan and Ordway 1966.

Allen, Thomas Barton. *The Quest: A Report on Extraterrestrial Life*. Philadelphia: Chilton Books, 1965. (An imaginative survey of the issues.)

Anderson, Paul. *Is There Life on Other Worlds?* New York and London: Collier-Macmillan, 1963. (Chapter 8, "On the Nature and Origin of Science," affords many perceptive observations.)

Ball, John A. "The Zoo Hypothesis," *Icarus* 19 (1973): 347–49. (Aliens are absent because the Intergalactic Council has designated Earth a nature reserve.)

———"Extraterrestrial Intelligence: Where Is Everybody?" *American Scientist* 68 (1980): 565–663.

Beck, Lewis White. "Extraterrestrial Intelligent Life," *Proceedings and Addresses of the American Philosophical Association* 45 (1971–72): 5–21. (A thoughtful and very learned discussion.)

Berrill, N. J. *Worlds Without End.* London: Macmillan, 1964. (A popular treatment.)

Bracewell, Ronald, N. *The Galactic Club: Intelligent Life in Outer Space.* San Francisco: W. H. Freeman, 1975. (A lively and enthusiastic survey of the issues.)

Breuer, Reinhard. *Contact With the Stars.* Trans. C. Payne-Gaposchkin and M. Lowery. New York: W. H. Freeman, 1982. (Maintains that we are the only technologically developed civilization in the galaxy.)

Cameron, A. G. W., ed. *Interstellar Communication: A Collection of Reprints and Original Contributions.* New York and Amsterdam: W. A. Benjamin, 1963. (A now somewhat dated but still useful collection.)

Dick, Steven J. *Plurality of Worlds: The Origins of the Extraterrestrial Life Debate from Democritus to Kant.* Cambridge: Cambridge University Press, 1982. (A lively and informative survey of the historical background.)

Dole, Stephen H. *Habitable Planets for Man.* New York: Blaisdell, 1964; 2d. ed., New York: American Elsevier, 1970. (A painstaking and sophisticated discussion.) A more popular version is S. H. Dole and Isaac Asimov, *Planets for Man.* New York: Random House, 1964.

Drake, Frank D. *Intelligent Life in Space.* New York and London: Macmillan, 1962. (A clearly written, popular account.)

Ehrensvaerd, Goesta. *Man on Another World.* Chicago and London: University of Chicago Press, 1965. (See especially chapter 10 on "Advanced Consciousness.")

Firsoff, V. A. *Life Beyond the Earth: A Study in Exobiology.* New York: Basic Books, 1963. (A detailed study of the biochemical possibilities for extraterrestrial life.)

Gavvay, Allen. "Les Principes foundamenteaus de la conaissance: Le Modele des intelligenes extraterrêtres." *Science, Histoire, Épistémologie: Actes du Premier Colloque Européen d'Histoire et Philosophie des Sciences.* Paris: J. Vrin, 1981, p. 33–59. (A stimulating philosophical discussion.)

Hart, M. H. "An Explanation for the Absence of Extraterrestrials on Earth," *Quarterly Journal of the Royal Astronomical Society* 16 (1975): 128–35. (A perceptive survey of this question.)

Herrmann, Joachim. *Leven auf anderen Sternen.* Guetersloh: Bertelsmann Verlag, 1963. (A thoughtful and comprehensive survey with special focus on the astronomical issues.)

von Hoerner, Sevastian. "Astronomical Aspects of Interstellar Communication," *Astronautica Acta* 18 (1973): 421–29. (A useful overview of key issues.)

Hoyle, Fred. *Of Men and Galaxies.* Seattle: University of Washington Press, 1966. (Speculations by one of the leading astrophysicists of the day.)

Huang, Su-Shu. "Life Outside the Solar System," *Scientific American* 202, no. 4 (April 1960): 55–63. (A useful discussion of some of the astrophysical issues.)

Huygens, Christiaan. *Cosmotheoros: The Celestial Worlds Discovered—New Conjectures Concerning the Planetary Worlds, Their Inhabitants and Productions.* London, 1698; reprinted London: F. Cass & Co., 1968. (A classic from another age.) Cf. Dick 1982.

Jeans, Sir James. "Is There Life in Other Worlds?" A 1941 Royal Institution lecture reprinted in H. Shapley et al., eds., *Readings in the Physical Sciences.* New York: Appleton-Century-Crofts, 1948, p. 112–17. (A stimulating analysis.)

Kaplan, S. A., ed. *Extraterrestrial Civilization: Problems of Interstellar Communication.* Jerusalem: Israel Program for Scientific Translations, 1971. (A collection of Russian scientific papers that present interesting theoretical work.)

Lem, S. *Summa Technologiae.* Krakow: Wyd. Lt., 1964. (To judge from the ample account given in Kaplan 1971, this book contains an extremely perceptive treatment of theoretical issues regarding extraterrestrial civilization. I have not, however, been able to consult the book itself.)

MacGowan, Roger A., and Frederick I. Ordway III. *Intelligence in the Universe.* Englewood Cliffs, N.J.: Prentice Hall, 1966. (A careful and informative survey of a wide range of relevant issues.)

——— "On the Possibilities of the Existence of Extraterrestrial Intelligence," in F. I. Ordway, ed., *Advances in Space Science and Technology.* New York and London: Academic Press, 1962, 4:39–111.

Nozick, Robert. "R.S.V.P.—A Story," *Commentary* 53 (1972): 66–68. (Perhaps letting aliens know about us is just too dangerous.)

Pucetti, Roland. *Persons: A Study of Possible Moral Agents in the Universe.* New York: Herder and Herder, 1969. (A stimulating philosophical treatment.) But see the sharply critical review by Ernan McMullin in *Icarus* 14 (1971): 291–94.

Rood, Robert T., and James S. Trefil. *Are We Alone: The Possibility of Extraterrestrial Civilization.* New York: Scribner's, 1981. (An interesting discussion of the key issues.)

Sagan, Carl. *The Cosmic Connection.* New York: Doubleday, 1973. (A well-written, popularly oriented account.)

——— *Cosmos.* New York: Random House, 1980. (A modern classic.)

Shapley, Harlow. *Of Stars and Men.* Boston: Beacon Press, 1958. (See especially the chapter entitled "An Inquiry Concerning Other Worlds.")

Shklovskii, I. S., and Carl Sagan. *Intelligent Life in the Universe.* San Francisco,

London, Amsterdam: Holden-Day, 1966. (A well-informed and provocative survey of the issues.)

Simpson, George Gaylord. "The Nonprevalence of Humanoids," *Science* 143 (1964): 769–75. Chapter 13 of *This View of Life: The World of an Evolutionist*. New York: Harcourt Brace, 1964. (An insightful account of the contingencies of evolutionary development by a master of the subject.)

Sullivan, Walter. *We Are Not Alone*. New York: McGraw Hill, 1964, rev. ed. 1965. (A very well-written survey of the historical background and the scientific issues.)

14. The Limits of Quantification in Human Affairs

Synopsis
(1) While science is often correlated with quantification, this is actually a very questionable idea. (2) Quantities have to be of a very special sort to constitute meaningful measures of something. (3 & 4) Various sorts of supposed "measures" are effectively meaningless. (5) In this domain a whole host of quantification fallacies arise. (6) Quantification does not provide an automatic entryway into the realm of scientifically meaningful measurement.

1. The Problem

Ever since the days of Britain's Factory Act inspectors of the last century, people have endeavored to base social planning on scientific principles. They have insisted that quantification is the name of the game here, but behind the facade of exactness and precision there lurk some big problems. As they see it, that which we cannot measure is thereby something we cannot adequately grasp at all.

Notwithstanding an extensive literature on the subject, the existing state of the art is such that our understanding of exactly what measurement is all about still leaves much to be desired.[1] The present discussion focuses on the problem of distinguishing genuine measurements—quantity-specifications that are informatively significant and meaningful—from those quantifications that lack this sort of substance. Its problem domain is that of the question: what ventures in quantification are not just feckless number-juggling but actually *measure* something?

Many treatises on the subject routinely suppose that each and every observationally based quantification represents a measurement.[2] But this is nonsense.

Quantification as such does not automatically constitute measurement. To see this, consider some examples of effectively meaningless quantities:
- the number of 3's in the distance (in kilometers) between two cities
- the number of times (on average) that the sentences of a given English text end with a proper name

It is very doubtful that such numbers do any measuring. For surely not every quantity represents a measurement. "How many of those girls remind you of your mother?" you ask me. "Two," I respond. A lovely quantity, that! But what in heaven's name am I *measuring?* It is thus all too clear that we must reject the claim of S. S. Stevens that "measurement [is] the assignment of numerals to objects or events according to a rule—any rule."[3]

Philosophers of science all too often sidestep the problem. They incline to ease their task by simply supposing that the problem of distinguishing between meaningless quantities and actual measurements will go away if we simply ignore it. But this tactic, however convenient, is very questionable. A measurement, after all, has to be a quantitative characterization of some meaningfully descriptive facet of reality, as opposed to one that is arbitrary and uninformative.

Exactly what is at issue here? A well-known pair of philosophical authors has told us that measurement consists in "indicating the quantitative relations between [intensive] qualities."[4] However, this idea that quantitative measurements must represent actually qualitative features of things is deeply problematic. Presumably the rate of exchange between dollars and yen constitutes some sort of measurement, but it is far from clear that the qualities or attributes of anything are at issue. Birth rates or inflation rates do not discernibly reflect qualities of anything—unless we blatantly *invent* something (a social or economic or political *system*) for them to be qualities of.[5] We can measure the annual snowfall of a place, but it is far from evident that one of its *qualities* is at stake.

Perhaps, then, one should abandon any reference to qualities in this context. Perhaps getting numbers is all that counts for measurement. Following in the footsteps of the operationalist school of P. W. Bridgman, the British philosopher of science Herbert Dingle has insisted that "instead of supposing a pre-existing 'property' which our operation measures, we should begin with our operation and its result, and then if we wish to speak of a property (which I do not think that we should do) define it in terms of that."[6] But here we are caught in a dilemma. If we tie measurement to specifically physical processes of quantity determination—linking it to apparatus manipulation and instrument pointer readings as with the measurement of length or mass—then we proceed in so restrictive a way that we have difficulties accommodating the sorts of quantities at issue with social affairs. Interest rates or the velocity of money circulation are some examples. In macroeconomics, after all, we get our quantities not by reading off the position of a pointer from a scale, but by copying suitably related numbers from pieces of paper. And yet the claims of these quantities to count as measurements seems to be conceded on all sides. It is thus very problematic to insist that measurements have to result from a *physical* measuring process of some sort. If, on the other hand, we loosen up the linkage of *measuring* to *measurement* too much, we lose our cognitive hold on what measurement is. In particular, if we reject the distinction between genuine *measurements* and merely *meaningless* quantities altogether—if the phenomenology we take into view is that of indiscriminate number-assignments—then there at once ceases to be any real reason why anyone should be interested in the topic of measurement at this absurdly general and undiscriminating level.

But how are we to understand this difference between real measurements and meaningless quantities? It would be all very well to say "I know a measurement when I see one," but this convenient approach facilitates understanding no more in the present context than in any other. In the absence of a sensible answer to the question of what it is to measure, there is clearly a large hole in our understanding of the scientific enterprise.

2. Quantification Versus Measurement: What Makes a Number Meaningful?

One sensible idea, it would seem, is that the only sort of quantity specification that deserves the name of measurement is that which plays an *ampliative*, information-expansive role in enabling us to extend our information regarding the items whose quantitative aspects are at issue. Thus contrast the preceding examples of problematic number allocation with such items as:

• inches of rainfall per annum (at a particular place)
• the number of inhabitants (of a particular town)

The situation here is very different. The salient difference seems to be one of cognitive utility. Information of this sort is genuinely *informative*. Given our knowledge of how things work in the world, we can draw various informative conclusions from it—regarding agricultural potential in the case of rainfall (say), and requirements for food and water in the case of population.

On this basis, the meaningfulness of quantities is clearly a reflection of the extent to which they are bearers of information. And it is evident that different, albeit interrelated considerations are at work in making a number meaningful in this regard. An adequate inventory is bound to include the following:

A. Effective Determinability

An actual measurement has to be the result of a practicable and implementable process. This is simply lacking in various cases. Consider, for example:

- the number of Latin words whose meaning a person once knew but has forgotten.

Or again, for living individuals contrast years from birth with years to death—that is, with

- a person's *remaining* life span.

The problem with such quantities is that they are effectively impossible to determine. We can make sense of the relative frequency with which a certain word is used by the writers of a certain language or the number of times their written discussions refer to a certain person, but there is little if anything we can do with

- the percentage of persons to whom a certain idea has occurred.

We cannot plausibly construe such problem quantities as meaningful measurements. Clearly, measurement has to be an operationally implementable procedure of some sort.

B. Reproducibility

An actual measurement has to be well defined—the stable result of a reproducible process and procedure. Real measurements must yield essentially the same result on different occasions when carried out by the same operator and must also yield essentially the same results when carried out concurrently by different operators. This requirement is illustrated by contrasting the number of "wrongdoers" with "the number of people found guilty of specific offenses (from some specified list)." Or again, contrast the number of unhappy people in a community with the number of suicides or the extent of its purchase of headache remedies.

C. Context Invariance/Robustness

A genuine *measurement* should reflect a substantially context-invariant quantity. Since talent and skill for taste-performance is generally comparative and context variable, industrial statistics regarding "skilled" versus "unskilled" workers are automatically suspect—particularly because the same product can be made from the same materials by processes and methods requiring very different sorts of talents.

In particular a genuine measurement must not be sensitive to factors that—as best we can tell—are not causally linked to the sort of thing at issue. For example, it just does not—and should not—matter whether we measure the amount of rainfall at a certain place on an odd or an even-numbered day, a workday or a holiday. In quantifying highly context-dependent relational and interactional aspects of things—say hair vibration as opposed to pulse rate—we are not in general carrying out actual *measurements*. (Hair vibration will vary haphazardly with wind exposure, body movements, head scratching, and so on.)

D. Validity/Coherence

With genuine measurements, there must be good reason to think that the result of the measurement is the true value of the quantity at issue. And this is only possible where different measurements cohere among themselves. Wherever there are different measuring processes and procedures—different ways of measuring what is supposed to be the same item—they must agree. Thus whenever the operators and their mode of operation themselves enter into the measurement process in a result-influencing way (as some physicists claim for the "measurements" of quantum phenomena, but as certainly holds for various sociological quantifications), the claims of the resulting numbers to represent measurements are ipso facto compromised.

E. Predictive Utility

It is clear that the usefulness of a number in predictive contexts is a pivotal feature of its meaningfulness. If we know a town's population we can use this datum to say something about this locality's consumption of water or its need for housing, while the age of its mayor or the average velocity of its winds from the east are, by comparison at least, meaningless quantities because few if any other quantities relevant to the constitution of towns can be predicted on the basis of these data.

This factor of the predictive serviceability of measurement is closely linked to yet another crucial aspect of measurement, namely:

F. Nomic Involvement

Measurements should result in quantities that function informatively in the context of general laws or lawlike statistical relationships.[7] This means, in particular, that, while any mathematical compounding of quantities will of course yield yet another quantity, there is no cogent reason for thinking that this is so with measurements. Even when the input quantities actually measure something, the functionally combined output quantities need not do so as well, owing to an absence of informative interrelationships. We would thus be highly disinclined to see

$$\sqrt{\text{age (in years)} + \text{weight (in kg.)}}$$

as *measuring* some aspect of a person's makeup. Given this quantity, there is effectively nothing else of any interest with which it stands in lawful interrelationships.

It is tempting to distinguish between "fundamental" measurements that we carry out "directly" by some suitably contrived process of physical manipulation with material objects and "derived" measurements that we effect by *calculations* with numbers resulting from prior measurements.[8] But—as the previous examples show—even when we put perfectly good measurements to work, it is clear that their arbitrary arithmetical compounding does not automatically measure anything. And this circumstance blocks the prospect of any *recursive* approach to the specification of legitimate measurements.

In this light, it is the laws of nature that ultimately determine the meaningfulness of complex quantities. Consider the gas law to the effect that (at least approximately) the product of pressure and volume divided by temperature is constant:

$$\frac{P \times V}{T} = \text{constant}$$

Given this law, the product $P \times V$ represents a perfectly meaningful quantity that provides us with a means of measuring temperature. But the quantity

$$P \times T \approx \frac{PT^2}{T} \approx T^2 \div V$$

fails to measure anything because it lacks any plausible physical meaning owing to its failure to function as a significant whole in the context of serviceable laws. As this example shows, when we combine perfectly meaningful and measurable quantities by throwing them together arbitrarily via rules of calculation, the upshot may well be something altogether nonsensical. The result of a real measurement has to be something that functions as a meaningful unit in the context of laws.

Genuine measurements accordingly have to be theory-laden: when we cannot embed quantities in a theoretical context of some sort and fit their results into our theories, then these number specifications are clearly unable to qualify as authentic measurements.[9] And if the theories into which they fit are patently invalid, absurd, or crazy, then the quantities at issue cannot qualify as genuine measurements. The range of evil-eye potency or the intensity of a person's "spiritual aura" are putative measures that vanish into nothingness with the disappearance of the theories on which they are predicated.

On the other hand, the mere fact that quantities do indeed behave in lawful coordination is of itself not enough to mean that they represent meaningful measurements. Thus consider once again the gas law:

$$P \times V \approx T$$

And let

$$P^* = P \times M$$

and

$$T^* = T \times M$$

where M represents some otherwise irrelevant quantitative feature of the system at issue—its aggregate mass—or, if you prefer, the time of sunrise in its location. Then those star quantities are—by their very design—merely rubbish. Nevertheless, they stand in splendid lawful coordination via the gas law. As this example indicates, the law-connectedness of determinable quantities does not of itself suffice to establish them as meaningful measurements.

We arrive at the recognition that each of the six measurement-characterizing factors considered individually—determinability, reproducibility, robustness, nomicity, and the rest—has its problems. More ominously yet, it emerges that, even in combination, they do not suffice to assure meaningfulness. They are all *necessary* for genuine measurement, but even jointly they are still not *sufficient*. Consider, for example, the following person-descriptive parameter:

- the sum of a person's age in years plus their systolic blood pressure.

Not only is this well-defined, determinable, and contextually stable, but it also has nomic involvement and predictive value in relation to such issues as general health and life expectancy. All the same, it is not something we could regard as a meaningful measurement because of the rather eccentric way in which it mixes apples and oranges. As such considerations indicate, something more is needed—over and above all of the preceding factors—to establish the claims of a quality to represent a meaningful measurement. A clearly appropriate further step moves in the direction of:

G. Dimensionality

Actually to *measure* something is to affix a numerical yardstick to some quantitative *parameter* that has its operative foothold in the world's scheme of things—length, temperature, mass, electric charge, money circulation, or the like. Measurement must present an objective and well-defined quantitative aspect of the qualitative makeup of things in the real world. A measurement, after all, is a number assignment made under a particular description such as:

- is x inches long
- measures x units on the Richter scale

- rates *x* units of acquisitiveness on the Spencer scale.

When numerical assignments fail to capture such lawfully descriptive features of things, then they just do not *measure* anything.

With a genuine measurement two questions arise:

(1) *What* is it that one is measuring? (The *object* question.)

(2) *How* is it that one is measuring what is at issue? (The *process* question.)

Moreover, the issues involved in (1) and (2) must be separable and distinct—that is, one must in principle be able to provide an answer to the *what* question independently of one's answer to the *how* question. Thus a person's weight or age represent cogent measurements, but not:

- age in years divided by waistline in inches
- height in feet plus years resided at present address

It is clear in this context that a wide roadway leading from meaningful to meaningless numbers is provided by mathematical compounding. This is indicated by such examples as:

- the product of the longest and shortest side of a polygon
- weight of a person in kg minus months resided at present address
- volume of an object (in cc^3) divided by its age (in years).

Such amalgamations are problematic precisely because there is no noncircular reply to the question: Just what is it that is being purportedly measured in this way? That is, we have no workable way of distinguishing the *what* of the putative measurement from its *how*.

The issue of dimensionality is clearly serviceable in providing a cogent factor that disqualifies such Rube Goldberg quantities from counting as genuine measurements. Regrettably, however, it is not easy to say just what is at issue here (which may explain why recent treatments of measurement in general simply bypass this issue of descriptive dimensionality). Philosophers of science have in recent years deliberated a good deal about what is or is not a *natural kind* when it comes to *classification*, but they have largely ignored the closely parallel and inherently no less important issue of what constitutes a *natural dimension* in point of *mensuration*.[10] Yet however difficult this issue may be to resolve, it is clear that such descriptive dimensionality—or something very like it—must be added to our list of requisites.

Our deliberations have indicated that the following features all represent necessary conditions for meaningful measurements.

- effective determinability
- reproducibility

- context stability/robustness
- coherence/validity
- nomic involvement
- predictive utility
- descriptive dimensionality

The (yet unresolved) problem we face is whether this collection of severally necessary conditions for genuine measurement is jointly sufficient. One may be tempted to surmise that this is indeed the case, but there is good reason to think that it is not. For one thing, even perfectly meaningful quantities can be contextually problematic. It makes sense to ask for the market price of gold or of lead specifically but not of metal in general. We can make good sense of the idea of the average birth-weight of a human female in specific but can make little of the color of the average death-weight of a fish in general. An even graver difficulty lurks around the corner. Consider, for the sake of an example, the person-correlative quantity:

$$\text{fage} = \text{age} \times \frac{\text{\# of fingers}}{\text{\# of toes}}$$

This clearly quantifies an aspect of a person's individual makeup in a way that (1) has a meaningful dimension (viz. chronological age, seeing that the fractional multiplier is dimensionless), and (2) behaves lawfully since for the vast majority of humans *fage = age*, and a person's age factors lawfully in many contexts and as biological development and life expectancy. Despite all this, however, the Rube Goldberg nature of the conception would leave one disinclined to consider "fage"-determination as a meaningful measurement process. What we want and need is a more cogent and discriminating account. But it is one thing to ask for such an account and another to know where it can be found.

Thus, in the end, it appears that we are confronted with a distinctly discomfiting situation. It is clear *that* we need a cogent way of distinguishing between meaningless quantities and genuine measurements, but it is far from clear *how* we are to draw this distinction.

In the philosophy of science, there are a number of structurally analogous problems of distinction and demarcation—between natural kinds and haphazard collections, between genuine laws and accidental generalizations, between ad hoc explanations and genuine ones, and even between real science and pseudoscience. The problem of distinguishing between mere quantification and real measurement is of this same general sort. But unlike the rest, it is a problem that has received precious little attention in the literature of the field. In this case even more than with others we have to do with an instrumentality of communication about science that is clearly useful, but which, nevertheless, we are simply unable to render clear and precise in anything like a satisfactory way. On all indications, the distinction between actual measurements and meaningless qualities remains enigmatic.

Some numbers can be acknowledged as measurements because, like weight and distance, they are paradigmatic of the very concept. Others are clearly not measurements because they violate one or another of the necessary conditions of the conception. But there is a considerable gray area where we do not see the way clear, and where we have good reason for caution and unease. And nowhere does this lack of a clear-cut way of drawing the line manifest itself more painfully than in the social sciences. Let us explore some of the ramifications of this situation.

3. Problematic Measurements

The need for maintaining the distinction between *measurement* and mere *quantification*, along the lines of the preceding deliberations, raises some disturbing questions regarding the status of various putative measurements that people nowadays conjure within the social sciences—particularly in areas that have implications for and applications to matters of practical policy. In economics, for example, we deal with quantities—such as interest rates—that are neither determined by actually measuring anything, nor yet by calculation from such quantities, but which emerge from calculations with numbers that crop up—mysteriously or otherwise—as writing on bits of paper. The idea of a meaningfully descriptive dimensionality in this domain is particularly problematic with respect to many of the quantities used in social analyses and social studies, where people throw numbers together in ways that ought to raise a theoretician's hackles. A good illustration of such a problem number is provided by the IQ tester's idea of a pervasive "intelligence." How cognitively competent people are is clearly of interest to us in many contexts. For, all too evidently, people have very different levels of ability—be it bodily or mental. But there is no good reason to expect that someone's physical or intellectual dexterity or versatility can be represented by a single number. The fact that it would be *convenient* if this was so from the angle of our programs of educational planning and management just does not make it true. Intelligence quotients throw together ability to decode language with analogy spotting and competence at calculation. Those tests explore very different sorts of abilities, and there is no reason to think that these can simply be aggregated by weighted addition to yield a meaningful composite result. Even a student's performance on one of those objective tests, on which we Americans place so much weight in college admissions processing and in scholarship awards, actually indicates little about ability for intellectually demanding work—or for doing anything substantially different from taking objective tests.

Of course, calculating with such quantities poses further problems. As our preceding deliberations have shown all too clearly, when one combines different quantitative factors by mathematical compounding one will generally get not a measurement but a mess. While this mess may possibly behave lawfully in some respects, nevertheless from the angle of *understanding*—of facilitating a conceptualized grasp of things—it remains a mess unless and until it comes to play a substantial role in a systematic law-framework of theoretical systematization.

4. Quality of Life as an Example

For another example of problematic qualification in the social domain, consider the nowadays fashionable quality of life measures (QOL) for different towns, areas, or countries. Here income data, housing costs, crime statistics, and cultural facilities are somehow quantified and then thrown together in some sort of aggregation. But there is clearly no good reason to think that a meaningful result emerges. Conceivably such QOL figures could be used to predict migration patterns, although this has apparently never been attempted. ("Other things equal, people abandon low QOL areas for high QOL ones" is hard to test and unlikely to survive testing in unqualified generality.) Clearly some embedding in a structure of lawful order is needed and has never been supplied. And there is an even more fundamental problem here.

In general, things have many different value aspects and we have no workable single-valued "function of combination" enabling us to extract a single, all-embracing *measure* of overall value from them. We have to deal with a plurality of distinct considerations that are not easily weighed off against one another. Even if we assume (perhaps rashly) that measurability is possible *within* each of these parametric dimensions, there is generally no way of making quantitative exchanges across different value parameters by way of weighing them off against each other in a common scale. The elements of a good journey are clearly not commensurable: adding more spectacular scenery cannot make up for bad food, even as adding more salt, no matter how much, will not compensate for a cake's lack of sugar. We can readily quantify various features of urban life—crime rates, housing costs, park acreage per capita, and so on. But does that mean we can combine such numbers into one single meaningful index of quality of life? Surely our prospects in this direction are no better than those of the nineteenth-century French physicians who thought that "health" was a property of the human body and attempted to provide a measure of a patient's "health" or "constitution" by means of a single number concocted out of a witches' brew of diverse factors.[11] With quality of life, or health or other such factors, the various evaluative aspects of the good are not interchangeable. In general, when we have to appraise a profile or complex of desiderata, preferability becomes a matter of a contextually determined *structural harmonization* rather than of mensurational maximization.

The idea of the social value of a life affords yet another example of a deeply problematic quantity. We shall, however, defer this issue for the present because it is addressed in the next chapter.

5. Fallacies of Quantification

So far we have been concerned to argue that it is wrong to maintain that whenever one can quantify one will obtain cognitively meaningful measurements. The other side of the coin also deserves examination. Is it correct to argue: whenever one *cannot* quantify one cannot accomplish cognitively significant work? Our deliberations about this issue will consider four widely current "fallacies of quantification" and their implications.

The first of these fallacies is that of the thesis:

Quantification and measurement are one and the same.

This is a fallacy we have already considered at length.

Another fallacy that deserves scrutiny is:

What one cannot quantify is not important.

The modern mania for numbers is notorious (especially in the United States!). Statistics have become a latter-day Baconian idol of the tribe. We love to rate and rank—everything from the world's best tennis player (even though performance varies substantially with playing surface) to the world's favorite soft drink or airline company. We worship at the altar of statistics: the thirst for measurements and quantities is a salient characteristic of our culture. Living in a society and in an era that is bedazzled by quantification, we have become *Homo numerans*, quantifying man. The splendid successes of natural science have enticed even the least numerate of us to the tempting idea that measurement and quantification afford the only true pathway and genuine understanding. Modern bureaucracy's commerce in paperwork and reports is just one more illustration of a dedication to statistics—not necessarily with a view to any pressing particular need, but nowadays often seemingly for their own sake.

It is this idolatry of numbers that underlies the great characteristic delusion of the times represented by the aforementioned fallacy. A laudable impetus to quantify the things that are important tempts us into the folly of deeming the things we cannot quantify to be negligible.

Testifying before the Presidential Inquiry Commission on the explosion of the shuttle *Challenger*, the representative of Morton Thiokol, the contractor that produced the booster rockets, explained their frustrations in persuading their NASA counterparts that it was unsafe to launch in the low-temperature conditions of the morning of January 26, 1986. Thiokol's representative felt that "we had to prove to them (NASA) that we weren't ready to fly." This was difficult to do because, when it came to the effect of cold on the seals that were at the heart of the problem, "I recognized that it's very difficult to quite quantify at which temperature these seals may be acceptable and at which they aren't acceptable. Now based on that data—some of it certainly was inconclusive—there was no doubt in my mind. And that's a difficult thing to quantify."[12] Given this circumstance, NASA refused to be convinced. There was little doubt in the mind of this witness that some shortcomings of quantifiability led to the discounting of his data.

This approach is, quite clearly, the height of folly. "What you can't quantify doesn't matter" is about as silly a thesis as "What you don't see can't hurt you." This quantification fallacy is just that—a fallacy. That night's debate in Florida between Thiokol and NASA puts before us a vivid and terrible illustration of the quantification fallacy at work.

The converse of the present fallacy, the idea that *whatever can be quantified is therefore important,* also has its problems. It too is clearly a fallacy, given the wide scope there is for quantitative trivia. It is easy to fool ourselves, however, into thinking that the things we can readily measure are the ones that count.

Numbers do not always tell the story. The wise strategist realizes that God is not always on the side of the big battalions. "How many divisions has the Pope?" Stalin asked. It is not a question that bemused General Jaruzelsky in Poland. By all quantitative yardsticks the United States was ahead throughout the Vietnam conflict and duly won the battles—but lost the war. The numbers just do not always reflect conveniently the impact things will eventually exert upon the world's course. In complex situations, the quantitative factors may be the easiest to get hold of, but they are not necessarily the most pivotal. Certainties are more easily measured than uncertainties, simplicities than complexities, but they are not necessarily the determining factors.

Our sense of legitimacy no longer relies on perceived quality or reputation but demands validation through numerical rating-indices. We demand quantitative performance ratings for our products, livability ratings for our cities, and quality indices for our schools. We rank colleges not on the competence of their graduates but in terms of "quality point averages" and scores on "objective" examinations. With tennis players performance on the courts is not an end in itself but a mere means to good computer rankings.

Time and again we need to be reminded of the importance of a concern for the things we cannot quantify—or cannot yet quantify in the existing state of information. Public opinion polls or the statistical content analysis of newspaper articles in the local press may have provided hard data but certainly were of little help to someone trying to judge how well-entrenched or vulnerable the position of the Shah was in Iran. To judge by the coverage of social issues in the media, thoughtful policy debate is impossible without a reliance on statistics. Common sense and historical analogy count for little. And yet numbers too can speak with a forked tongue. (Extrapolation is a dangerous sport; be it with hemlines or employment statistics, a 1 percent shift this year seldom presages a 10 percent shift a decade further on.)

We forget all too easily that there are "lies, damned lies, and statistics," as Benjamin Disraeli put it. The numbers are not a substitute for sound judgment. On the contrary, it is generally only on the basis of sound judgment that they themselves are useful. Admittedly, the sensible appraisal of situations is often encouraged and facilitated by numerical data; but statistics only carry a meaningful message to the prepared mind.

There is no need to deny the plain fact that quantitative information is often illuminating and useful. The point, rather, is that there is no earthly reason to think some sort of correlation exists between the ease and accuracy of the quantifiability of some consideration and the significance of its role in the matters to which it relates. The remedy of the fallacy at issue lies in the recognizing that: *the things you cannot quantify in the context of an inquiry may well turn out to be the most important.* As long as we persist in adhering to the quantification fallacy by thinking that what one cannot quantify just is not important and can safely be ignored, we shall also persist in trying to send aloft rockets, economic policies, and social programs that just will not fly.

Let us now turn to our third fallacy, which is clearly related to the preceding one:

Everything (at any rate, everything that is even remotely interesting) can be quantified.

Fallacy no. 2 maintained that whatever is important can be quantified; the present fallacy says that whatever is really interesting can be.

A good illustration of this fallacy is provided by the IQ tester's idea of a pervasive "intelligence." To avert the fallacy at issue we must accordingly recognize that, despite what people may think to the contrary, *some highly significant and interesting matters simply cannot be quantified.*

Consideration of the preceding fallacies leads naturally toward the recognition of yet another:

The quantities that measure various distinct factors can always be combined in a meaningful aggregate. They can ultimately be seen as exchangeable parts of a composite whole.

This idea that different sorts of quantities can be lumped together as convertible constituents of one single composite—that they can be seen as exchangeable parts of an aggregate whole—runs deep in modern thought. All the same, it is highly unrealistic. Consider, for example, some of the points of merit that relate to the effectiveness of a car.

- maximum speed
- starting reliability
- operating reliability (freedom from breakage)
- passenger safety
- economy of operation

If the top speed of a car is 5 mph, no augmentation in passenger safety or operating reliability can make up for the shortcoming. Again, if the car is eminently unsafe, an increase in its other virtues cannot offset this. Where the various merits of a car are concerned, there simply is no free exchange among the relevant value parameters, but only complicated (nonlinear) trade-offs over a limited range. This illustrates a general situation.

In general, things have many different value aspects and we have no workable, single-valued "function of combination" enabling us to extract a single, all-embracing *measure* of overall value from them. We have to deal with a plurality of distinct parameters of value. Even if we assume (perhaps rashly) that measurability is possible *within* each of these parametric dimensions, there is generally no way of making quantitative exchanges across different value parameters by way of weighing them off against each other in a common scale. The elements of a good journey are not commensurable: adding more spectacular scenery cannot make up for bad food. We can readily quantify various features of urban life—crime rate, housing costs, park acreage per capita, and so on. But does that mean we can combine such numbers into one single meaningful index of quality of life?

To think (as per the economists' "utility") of assessing the preferability of goods by a single number is every bit as sensible (or foolish) as to think that we can reflect the intelligence of people in a single number—or even the value of a human life. What is going on throughout is an indulgence in number idolatry that rides roughshod over important differences and diversities.

The commitment to convertibility at issue in the economist's "utility" represents a harmless fiction in its proper sphere of a tradeable good whose value can be established in a common unit via a market exchange mechanism. To pretend to have a more general and pervasive instrument for the rational governance of human decisions is skating on very thin ice.

"The good" at large is something multidimensional and not homogeneous.

Enhancing it is a matter of optimizing a complex, not of maximizing a determinable quantity. The evaluative aspects of the goal are not interchangeable. We must harmonize rather than maximize. In general, when we have to appraise a profile or complex of desiderata, preferability becomes a matter of a contextually determined *structural harmonization* rather than of mensurational maximization.

We come thus to the need for recognizing that even when we are dealing with quantities that express meaningful measurements, the commensurability needed for an integrated measurement to result may not be possible. In many situations we may have to settle for quality-appreciative judgment where quantity-oriented measurement is simply infeasible.

Our cook's tour of quantificational fallacies is now completed. They represent principles of thought so deeply entrenched in our culture that it will seem to many to be problematic and tendentious even to suggest that they may be incorrect. However, seeing them as unwelcome does not, of course, make them any the less fallacious.

6. Larger Vistas

In the aggregate, then, these deliberations indicate that quantification is neither necessary nor sufficient to obtain cognitively meaningful measurements. Accordingly, they indicate the need for recognizing that, even when we are dealing with a perfectly fine *quantity*, the conditions needed for this to qualify as an authentic *measurement* may nevertheless not be satisfied. The importance of science in modern life has engendered a quantitative prejudice. People often incline to think that if something significant is to be said, then you can say it with numbers and thereby transmute it into a meaningful measurement. They are tempted to endorse Lord Kelvin's dictum that "When you cannot express it in numbers, your knowledge is of a meager and unsatisfactory kind."[13] When one looks at the issue more clearly and critically, however, one finds that there just is no convincing reason to think this is so on any universal and pervasive basis.

Science has succeeded in mathematicizing the realm of our *knowledge* to such an extent that we tend to lose sight of the fact that the realm of our *experience* is not all that congenial to measurement. It is full of colors, odors, and tastes, of likes and dislikes, of apprehensions and expectations, and so on, that are not particularly amenable to measurement. We readily forget how very special a situation actual *measurability* is—even in contexts of seeming precision.

Moreover, reliance on numbers brings in its wake a host of problems of its own. For one dangerous thing about numbers is that small errors in their use can produce large—and very unfortunate—consequences. A minor mistake in the number encoding of a prescription medication can prove lethal. For many years, spinach has enjoyed a great prestige as a valuable source of iron because a misplaced decimal point credited this vegetable with an iron content ten times its actual value.[14]

A fetish for quantification seems to be astir among our contemporaries. We worship at the altar of statistics: the penchant for quantities is a salient characteristic of contemporary western culture. Everything we touch turns to numbers: intelligence quotients, quality of life indices, feminine beauty (ranked on a scale from zero up to a "perfect 10"), and so on. It is thus easy to see why the prospect of meaningless quantities should cause unease. For in this measurement-enchanted time of ours we constantly invoke quantitative information as a basis for decision making and policy guidance. But garbage in, garbage out. If those quantities that people throw about so readily are in fact meaningless, then the decisions we so enthusiastically base upon them are built on sand. In many situations we would do well to settle for rough-and-ready judgmental appraisals because meaningful quantitative measurement is simply infeasible.

After all, numbers do not always tell the story. One does not need to enroll oneself, with Goethe, as an opponent to measurement's entry into the domain of human doings and dealings to feel a deep disquiet regarding the particular ways in which people have sought to introduce measurement into the social sphere. The fact is that in everyday life, professional practice, and public affairs alike, stubborn reliance on numbers can sometimes prove more of an obstacle than an aid to critical and reflective thought.

The immense success of quantitative techniques in the mathematicizing sciences has misled people into thinking that quantification is the only viable road to cogent information. But think—is it really so? Where is it written that numbers alone yield genuine understanding—that judgment based on structural analysis or qualitative harmonization is unhelpful and uninformative, so that where numbers cannot enter, intelligibility flies away? (After all, modern mathematics itself is not all that quantitative, seeing that it is deeply concerned with issues such as those of topology and group theory that deal with structures in a way that sets quantitative issues aside.)

It must be stressed, however, that to acknowledge the limits of measurability is not to downgrade the whole process, let alone to propose its abandonment. It is precisely because we are well advised to push the cause of measurement as far as we legitimately can that we need to be mindful of the line between meaningful measurements and meaningless quantifications. That we cannot draw this line better than seems to be the case at present is—or should be—a proper occasion for justified chagrin.

The hue and cry is audible in advance: "only through quantification and measurement are we put on the high road of a scientifically valid understanding of how things work in the world." Even to raise these questions is to be led toward a recognition that the idolatry of numbers is a prejudice or preconception that is not only questionable but potentially badly misleading. In everyday life, professional activity, and public affairs alike, reliance on numbers is no substitute for reflective thought. If science indeed is a matter of exact and rigorous thinking about the observably detectable ways of the world then it cannot be equivalent with quantification: mathematics then has to be seen as its servant and not as its master.[15]

15 The Limited Province of Natural Science

Synopsis

(1) One fundamental limitation attaches to natural science through the consideration that, after all, knowledge is only one human good among others, there being, in addition to specifically cognitive goods, also those relating to the quality of personal and communal life: to physical well-being, human companionship, environmental attractiveness, social harmoniousness, cultural development, and others. Man does not live by knowledge alone: there is a whole world of important human goods and goals outside the cognitive area. (2) Moreover, even in the specifically cognitive domain, scientific knowledge is only one particular sort of knowledge. Other cognitive projects exist in line with various different sorts of "understanding"—immediate experience, affective appreciation, and human sympathy and empathy, among others. (3) To be sure, these other enterprises are not in a position to compete with science within the range of its own proper concerns; their orientation is very different. Within its own cognitive province, science stands supreme. We have no place else to go: there is no alternative but to turn to the science of the day for whatever we want to know about the furnishings of the world and their modes of comportment. Whatever limitations and shortcomings science may have are simply beside the point: natural science is the only game in town that deals adequately with those issues that constitute its proper domain. (4) While science does indeed have various sorts of limitations, none of them affords grounds for complaint or reproach about the world of scientists as such—let alone for abandoning their project.

1. Knowledge as One Good Among Others

In former times, philosophers rightly objected to a naively anthropomorphized view of nature; nowadays they frequently object to an overly naturalized view of man—an exaggerated assimilation of the activities of man and of human society to processes of the developmentally prehuman domain. They often lament that the rationality of technical means and ends—the vaunted value *neutrality* of science—"liberates" us from human values and concerns in a way that sometimes threatens to dehumanize us by blunting our understanding and appreciation of characteristically human phenomena. To attack anthropomorphism is all well and good but has its limits when we come to man and his works.

Objection to an overly scientized view of man underlies much of the recent critique of the scientific outlook. Scientific "reason," we are told, is restricted to mere technical issues and fails to reckon appropriately with human *values*. Our scientific and technical intelligence embodies a rationality of process, rather than product, and tends to substitute deliberations about ways and means for an evaluative concern for outcomes and ultimate results.

But this sort of dichotomy is too facile. For one thing, science, like other human enterprises, is itself a locus of values—those relating to knowledge of and control over the course of nature's events, to cognitive and material goods. Moreover, the fact/norm distinction does not represent an absolute and impassable divide, because the pursuit of knowledge is itself governed by norms, with the conception of "making good a claim" figuring as a mediating link between the cognitive and nonnative.[1] Our factual contentions themselves rest on a right or entitlement of a certain sort—a right to maintain something under the aegis of epistemic ground rules. The pivotal consideration here is that of its being right, fitting, and proper for someone in certain epistemic circumstances to endorse a certain thesis. Rational acceptance is a cognitive act governed by appropriate nonnative standards, so that inquiry itself can, and should, be viewed as a mode of practical activity—as a cognitive praxis governed by norms and criteria. Indeed, even the issue of what is important *for science* is, to a significant degree, a matter of value considerations, and scientific progress can also be a matter of value-reorientation. (For example, the latter-day rise of cosmology reflects, in significant measure, a rethinking of what the important issues are.)

This said, however, it must be recognized that man does not live by knowledge alone. Other legitimate and important human enterprises exist and delimit the significance of science within the sphere of our concerns. While knowledge represents an important aspect of the good, it is by no means an aspect that is predominant and paramount. It is only one element of the constellation of human desiderata—one valuable project among others, whose cultivation is only one component of the wider framework of human purposes and interests. The quality of our lives turns on a broad spectrum of personal and communal desiderata such as physical well-being, human companionship, environmental attractiveness, social harmoniousness, cultural development, and so on—values toward whose attainment the insights afforded by science can often help us, but which themselves nevertheless fall outside its domain.

While the cultivation of knowledge is indeed only one worthy human project among many, it is, however, a particularly important one. Knowledge is a key component of the good per se, because of its smooth fit within the overall economy of norms. Its pursuit as a good in no way hinders the cultivation of other legitimate goods; on the contrary, it aids and facilitates their pursuit, thereby acquiring an *instrumental* value in addition to its value as an absolute good in its own right. Whatever other projects we may have in view—justice, health, environmental attractiveness, the cultivation of human relations, and so on—it is pretty much inevitable that their realization will be facilitated by the knowledge of relevant facts. Thus, even though the pursuit of knowledge is not our only appropriate task, it is nevertheless an enterprise whose normative standing is high because knowledge serves to facilitate the realization of any other legitimate good: any and every such good is cultivated the more effectively by someone who pursues its realization knowledgeably.

2. Scientific Knowledge as One Mode of Knowledge

Knowledge is only one human good among others, and its pursuit is only one valid objective among others. Moving beyond this point, it must also be recognized that, even in the strictly cognitive domain, scientific knowledge is only one sort of knowledge: there are other valid epistemic and intellectual projects apart from the scientific. The epistemic authority of science is great, but not all-inclusive.

Natural science is a mission-oriented endeavor, with its goal structure formed in terms of the traditional quartet of description, explanation, prediction, and control of nature. It inquires into what sort of things there are in the world and how they work at the level of law-governed generality, focusing primarily on the lawful modus operandi of the natural processes that characterize the furnishings of nature. Given this mission, the concern of science is, and must be, with the public face of things—with their *objective* facets. It strives for *reproducible* results, and its focus is on those objective features of things that *anybody* can discern (in suitable circumstances), regardless of his particular makeup or experiential background. Science deliberately sets aside the observer-relative dimension of experience. The English philosopher F. C. S.

Schiller put the key point with admirable clarity:

> Large tracts of actual experience are submerged and excluded as "subjective," in order to focus scientific attention upon the selected and preferred sections of experience which are judged fit to reveal objective reality.... Thus the differences between experienced particulars, even when not denied outright, are simply assumed to be irrelevant for scientific purposes, and are ignored as such. It is by this assumption alone that science is enabled to construct the common world of intersubjective intercourse, or "objective reality," which different observers can contrive to explore.[2]

The "facts" to which science addresses itself are accordingly those that arise from intersubjectively available observation rather than personal sensibility. Its data are those universals accessible to man qua man, rather than those that are in some degree subjective and personal—accessible only to people of this or that particular background or experiential conditioning. Thus, science ignores the individualized, affective, and person-linked dimension of human cognition: sympathy, empathy, feeling, insight, and personal reaction. The phenomena it takes as data for its theory projection and theory testing are publicly accessible. Value appreciation—how things strike people within the informational setting of their personal (and perhaps idiosyncratic) experiences or their sociocultural (group conditioned) background—is something science leaves aside; it concentrates on the impersonally measurable features of things. This quantitative orientation of our natural science means that the qualitative, affective, evaluative dimension of human cognition is bypassed. Our knowledge of the value dimension of experience—our recognition as such of these features of things in virtue of which we deem them beautiful or delightful or tragic—remains outside the range of science. Not experiential sensibility but theoretical understanding is the crux.

Accordingly, science omits from its register of facts that are worth taking at face value the whole affective, emotional, feeling-oriented side of our cognitive life. Such matters appear within science as explanatory problems rather than as building blocks for interpretation; they are seen as part of the problem rather than of the solution, as objects of study and explanation, not as data. The sort of observation-transcending experience that lies at the basis of norms and values (affective receptivity, for example) remains outside its range. Scientific awareness does not teach us to enjoy and appreciate. Issues of evaluation and appreciation are not matters to be settled by scientific inquiry into nature's ways, which teaches us the whys and wherefores of things but does not instruct us about their worth.

Thus, no matter how far we push science forward along the physical, chemical, biological, and psychological fronts, there are issues about humanity and its works that will remain intractable by scientific means—not because science is impotent within its range but because these issues lie outside it. We shall always have questions about humanity and its place in this world's scheme that lie outside the mandate of science.

Moreover, in natural science we investigate what is and what can be relative to the laws of nature. The more remote regions of possibility—the more imaginative realms of what might be and what might have been—lie outside our purview in natural science, which, after all, does not deal with purely imaginative constructions and conjectured possibilities. Virtually by definition, natural science is oriented toward what actually exists in nature and does not deal with the speculative domain of what is not but yet might be. Its concern is with the realm of actuality, the real; the realm of the imaginatively hypothetical is outside its sphere. The artistic, imaginative aspect of human creativity—the projection of abstract form in art or music—thus lies outside its range of concern. Poetry, drama, religion, proverbial wisdom, and so on all carry messages that cannot be conveyed within the medium of scientific discourse. The limits of science inhere in the limits of its cognitive mission and mandate: the "disinterested" depiction and rationalization of objective fact.

The "knowledge" with which science equips us is descriptive rather than normative, with the issue of evaluatively responsive appreciation—of what sorts of things are worthwhile—simply left aside. However value-laden the *pursuit of* scientific knowledge may be, on the side of *content* science remains value-neutral. Substantively considered, science is value free; its approach to the characterization of phenomena is simply devoid of the element of personalized evaluation. Natural science aims at mastery over nature—at controlling it both physically and intellectually. It abstracts from the enterprise of *appreciating* nature. The affective, appreciative, emotively evocative side of human knowledge, the intuitive and unreasoned ways of knowing—the mechanisms through which we standardly understand other people and their productions—lie outside science's range of concern. Yet, *feeling* too is a mode of cognition, although one that, in its affective impurity, falls outside the scope of science as insufficiently "objective."

Thanks to its exclusion of normative and evaluative issues, science approaches people as objects of study, as *things*, and not as *persons*. The sort of knowledge of a person in which it trades is not that of a friend or companion, based on concern inherent in mutual interaction, but the detached, impersonal knowledge of a doctor, biologist, or sociologist—the knowledge of observation rather than communion. The aspects of mutual recognition and interactive reciprocity are lost; and such cognitive modes as sympathy and empathy are put aside. The scientific mode of understanding takes the externalized route of causal explanation, and not the internalized route of affective interpretation.

The cognitive approach of science to man's understanding of man deliberately puts aside that element of recognition as a fellow man characteristic of all genuinely human relationships among people.

Many people sense a shudder of fear when science turns its cold, objective gaze upon man and his works—as though our deepest human values were somehow being put at risk. But nothing could be more foolish. To fear science as *antithetical* (rather than simply *indifferent*) to human values and interests is not only discomforting, it is irrational and inappropriate, because based on a profound misunderstanding of what science is all about. The fact that a human being is an agglomeration of chemicals, a complex of flesh and bone evolved from creatures of the primeval slime, is no reason why he cannot also be a friend. Objectively, the playing of a violin is no more than a scraping of catgut, the glow of a sunset no more than a shower of radiation. Does that prevent them from reaching heights of transcendent beauty? To speak of conflict here is to become enmeshed in a confusion between diverse perspectives of consideration—between different levels of thought and inquiry.[3] Science is not *opposed* to these concerns, it is irrelevant to them; it simply ignores them, having other fish to fry.

Accordingly, as the "idealist" tradition of German philosophy from Hegel to Heidegger[4] has always rightmindedly insisted, the quest for scientific knowledge is simply one human project among others. Reacting also against the Greek view that *episteme* is supreme, this tradition replaced the old theological doctrine that faith transcends knowledge with the latter-day variant that *sensibility* transcends knowledge. Such a position has much to recommend it.

Science, then, has its limitations. It is no be-all and end-all: it cannot answer all the questions that matter to us. But we are dealing here with the limitation of a range of concerns as such, and not with limitation of capacity within this range. The "limitations" at issue are simply mission-inherent. Science has a determinate mission of its own in rationalizing the objective empirical facts; and here, as elsewhere, determination is negation: because science is a certain definite kind of enterprise, there are also things that science is not. These other cognitive ventures are not *alternatives* to science, because their cognitive orientation lies toward other directions. They are not different ways of doing the same job but look to doing altogether different sorts of jobs. And so, these other cognitive projects are not in a position to compete with science in its own domain; in fact, there is no question of competition, because diverse objectives are at issue. Those who play different games are not in competition with each other.

It must thus be stressed that what we are dealing with here is something that is not a defect or a shortcoming. It is a disability imposed by the aims of the enterprise—the objectives that characterize science as the thing it is. The characteristic cognitive task of science is the *description* and *explanation* of the phenomena—the answering of our *how?* and *why?* questions about the workings of the world. Normative questions of value, significance, legitimacy, and the like are simply beside the point of this project. The fact that there are issues outside its domain is not a defect of natural science but an essential aspect of its nature as a particular enterprise with a mission of its own. It is no more a defect of science that it does not deal with belles lettres than it is a defect of dentistry that it does not deal with furniture repair. It is no deficiency of a screwdriver that it does not do the work of a hammer.

Science's domain-inherent disability in this regard of having a finite and determinate mission is something that the practitioners of science are well advised to keep in mind. Inflating the claims of science to the point where it is held to have all the answers about the condition of man, the meaning of life, or the objects of social polity is a dangerous move. For it is always ill advised to make exaggerated claims that go beyond what can be delivered. Such an inflated view of capacities invites skepticism and hostility in the wake of the disappointment of expectations that is its inevitable consequence. The upshot is bound to be a lamentable failure to secure proper credit for the inestimably great benefits we derive from science within its proper domain.

The theorist who maintains that science is the be-all and end-all—that what is not in science textbooks is not worth knowing—is an ideologist with a peculiar and distorted doctrine of his own. For him, science is no longer a sector of the cognitive enterprise but an all-inclusive worldview. This is the doctrine not of *science* but of *scientism*. To take this stance is not to celebrate science but to distort it by casting the mantle of its authority over issues it was never meant to address.

Ludwig Wittgenstein wrote that "We feel that even if *all possible* scientific questions be answered, the problems of life have still not been touched at all. Of course there is then no question left, and just this is the answer."[5] This austere perspective pivots on the view that scientific issues are the only ones there are—that where no scientific question is at issue, nothing remains to be said, and that factual information is the end of the cognitive line. If this position is adopted, then questions relating to normative and evaluative issues of significance, meaning, and validity—questions relative to beauty or duty or justice, for example—can all be set at naught. Such a response does indeed resolve the problems of life, but only by casting them away into the outer darkness. This scientific positivism is indeed antipathetic to human values. As one acute writer has observed:

> [Such a doctrine] is an attempt to consolidate science as a self-sufficient activity which exhausts all possible ways of appropriating the world intellectually. In this

radical positivist view the realities of the world—which can, of course, be interpreted by natural science, but which are in addition an object of man's extreme curiosity, a source of fear or disgust, an occasion for commitment or rejection—if they are to be encompassed by reflection and expressed in words, can be reduced to their empirical properties. Suffering, death, ideological conflict, social classes, antithetical values of any kind—all are declared out of bounds, matters we can only be silent about, in obedience to the principle of verifiability. Positivism so understood is an act of escape from commitments, an escape masked as a definition of knowledge.[6]

Nothing within or about science demands such a dehumanization of our sensibilities. To take this stance is not to celebrate science but to distort it.

Science has no concern with matters that do not come within our view of its problem mandate, the domain of natural fact. For example, science as such is not concerned with values as actually espoused, rather than held at arm's length as objects of descriptive scrutiny. But such domain restrictions resulting from a finitude of the problem mandate of science are hardly a basis for rejecting as nonsense what falls outside its sphere.

Science does not have exclusive rights to knowledge: its province is far narrower than that of inquiring reason in general. Even among the modes of knowledge, science represents only one among others. It is geared to the use of theory to triangulate from objectively observational experience to answer to our questions about how things work in the world. There are many other areas, however, in which we have a cognitive interest—areas wholly outside the province of science.

Man is a member not just of the *natural* but of the specifically *human* order of things. There is more to reality than science contemplates; in the harsh but stimulating school of life, we are set examinations involving problems for whose resolution our science courses by themselves do not equip us.

3. The Autonomy of Science

In bringing the discussion toward its close, let us return to our central question of the explanatory range of science. Is science so limited that certain factual questions regarding the nature of things simply lie beyond its powers?

The answer here is a clear negative: *nothing* in the way of factual issues about the world and its modus operandi is in principle to be placed outside the purview of natural science. The conceivable subjects of scientific explanation accordingly exhibit an enormous, indeed an endless, variety. All the properties and states of things, any and all occurrences and events, the behavior and doings of people—in short, every facet of what goes on in the world—can be regarded as appropriate objects of scientific scrutiny and explanation. No matter of natural fact lies in principle outside the domain.

The terrain of science is far from all-inclusive, but in its own province, science stands supreme. Paul Feyerabend maintains in *Against Method* that science is itself no more than one form of ideology and that we would be well advised to replace it with an anarchistic potpourri according to which "anything goes." The claims of science to provide objective and useful information about the world are, he argues, no better than those of myth or idle speculation.[7] But this sort of thing is easier said than substantiated. Where are those informative disciplines, rival to standard science, that embrace alternative modes of medicine, engineering, and so on capable of matching the applications of standard science in actual effectiveness? Where are the rival theory systems that can approach science in applicative and predictive power? The point, of course, is not that alternative belief structures about the world are theoretically impossible or evaluatively unappealing; it is simply that, in comparison to orthodox science, they are hopelessly ineffectual.

The situation is simple and straightforward. If we want to know about the constituents of this world and their laws of operation, we have to turn to science—and in fact to the science of the day. Whatever its shortcomings or limitations, science is the only game in town with respect to our best available picture of the laws of nature. There is no place else to turn for information worthy of our trust. (Tea-leaf reading, numerology, the Delphic oracle, and the like are not serious alternatives.) If we want to be informed about the furnishings of the world and their modes of comportment, there is simply nowhere else to go.

On its home ground, science has no effective challenge, no serious competition. Whatever its limitations and lack of completeness, natural science is wholly self-sufficient within its proper domain of the explanation and prediction of natural phenomena. And this is how the matter must stand, given the crucial fact that science is autonomous. Corrections to science must come from science. Science is inexorably complete in regard to its self-sufficiency. The key point was made long ago in Hegel's *Phenomenology of Spirit*.[8] We cannot set limits to our (scientific) knowledge of reality, because any relating of knowledge to reality has to come from within that knowledge itself. Inquiring thought cannot get outside itself to compare its deliverance with the real truth. There is no viable *external* standard by which the deliverances of science can be appraised. We have no choice but to follow science where it leads us. There is no other, science-external cognitive resource to monitor its operations. Shortcomings in scientific work can emerge only from further scientific work. The defects of science can only be removed by the further results of science.

The acceptability of scientific claims is a matter to be settled wholly at the level of considerations internal to the scientific enterprise. A "science" viewed as subject to external standards of correctness is simply not deserving of the name. Scientific claims must—whenever corrected at all—be corrected by further scientific claims. This fundamental fact is the rock bottom that provides the only basis on which the doctrine of the *self*-sufficiency of science can find its foothold.

Thus, while there are also other sectors of the cognitive sphere, and natural science is only one cognitive discipline among others, the fact remains that within its own domain it is sovereign. Within the sphere of its appropriate jurisdiction, so to speak, science is supreme because it stands unrivaled and alone. Whatever be its limitations, science is our only resource for dealing adequately with the issues that constitute its proper domain.

4. Conclusion

Returning to the start of our discussion, we may note that the preceding deliberations have led to four significant—and significantly different—conclusions:

1. Natural science has no limits (in the strict sense of that term). There is no reason to think, on the basis of general principles, that any issues within the domain of natural science lie beyond its capabilities (chapters 6 and 7).
2. However, there is every reason to think that (in consequence of Kant's Principle of Question Propagation) natural science can never be completed: it can never manage to resolve all of its questions. Natural science thus has domain-internal incapacities (chapters 2–4).
3. There are also powerful limitations to science of practical (and ultimately economic) provenience. One cannot, however, adduce any concrete examples of irresolvable insolubilia (chapters 8 and 9).
4. There is no question that natural science is subject to domain-external incapacities. We must recognize that various important evaluative and cognitive issues lie altogether outside the province of science as we know it (chapters 10 and 11).

These four points summarize the main results of our analysis.

The question remains whether the various disabilities of science are an appropriate subject of dismay. To address this question, we would do well to begin by distinguishing between the sorts of considerations that afford a basis for *lament* (or *regret*) and those that afford one for *complaint* (or *reproach*).

If something desirable cannot possibly be achieved on the basis of general principles inherent in the very nature of the case, then we might wistfully lament (or regret) this circumstance—that is, we can quite appropriately have a longing that it should be otherwise. But in this case we must also recognize that such regret succumbs to unrealism, because the situation we face is simply unalterable. Accordingly, we have no basis for complaint or reproach or recrimination of any sort regarding the incapacity at issue.

The distinction between the regrettable and the complaint-worthy bears importantly on our conclusions. For the disabilities at issue in the limits and limitations of science are not such that complaint or reproach is warranted. To see that this is so, we can proceed by viewing the matter either particularistically or holistically. Consider the two salient questions:

1. Is there any *particular issue* that natural science can resolve in theory, of which we can say with assurance that it cannot resolve this issue in practice?
2. Can we say of the *holistic state* of attaining completeness—of resolving *all* its issues—that natural science might achieve it in theory but not realize it in practice?

The answer to both questions is an emphatic no, but for very different reasons. With respect to question (1), we can never say with assurance that science cannot manage to resolve a certain issue. And with respect to question (2), we cannot claim that science might (even in theory) manage to achieve absolute completeness.

Accordingly, the limits and incapacities of science do not represent shortcomings that might be remedied but for some failing on our part. To be sure, there will *always* be attainable scientific findings that we do not have in hand. But this circumstance, too, is inevitable—inherent in the very structure of the enterprise. Thus, while science does indeed have various disabilities, they do not afford—be it singly or collectively—any reasonable grounds for discouragement, let alone for complaint or disillusionment about the enterprise as such.

NOTES

Introduction

1. However, see vol. 107 (1978) of *Daedalus*, which presents an interesting collection of papers dealing with some of these issues.
2. On these issues, see John Passmore's excellent book, *Science and Its Critics* (London: Duckworth, 1978).

Chapter 1

1. The question Q presupposes the proposition p (symbolically: $Q \ni p$) if every possible explicit answer to Q entails p:

    $Q \ni p$ iff $(\forall q) [p @ Q \supset (q \rightarrow p)]$.

    Here "$p @ Q$" stands for "the thesis p affords an explicit answer to the question Q." Following standard logicians' usage, "iff" stands for "if and only if" throughout and the arrow represents strict as opposed to merely material implication.

2. Symbolically:

    $Q \in \mathbf{Q}(S)$ iff $(\forall p) [Q \ni p \supset p \in S]$.

3. Symbolically:

    $Q \notin \mathbf{Q}(S)$ iff $(\exists p) [(Q \ni p) \, \& \, (p \notin S)]$.

4. Compare Adolf Grünbaum, "Can a Theory Answer More Questions than One of Its Rivals?" *British Journal for the Philosophy of Science*, vol. 27 (1976), 1–22.
5. Kant was perhaps the first philosopher to give serious attention to developing the theory of scientific questions and to exploit it as an instrument of epistemological method. See the author's paper on "Kant and the Epistemology of Scientific Questions," in Joachim Kopper et al., eds., *200 Jahre Kritik der reinen Vernunft* (Hildesheim: Gertsenberg, 1981), 313–34. However, Kant's initiative proved infertile, and the topic of questions long lay dormant until finally put on the agenda of twentieth-century philosophy by R. G. Collingwood. See his *Essays on Metaphysics* (Oxford: Clarendon Press, 1940).

6. Symbolically:

$$(\forall Q)(\forall t)(\forall t')\{[t < t' \ \& \ Q \in \mathbf{Q}(S_t)] \supset Q \in \mathbf{Q}(S_{t'}')\}$$

7. The progress of science offers innumerable illustrations of this phenomenon, as does the process of individual maturation:

   After three or thereabouts, the child begins asking himself and those around him questions, of which the most frequently noticed are the "why" questions. By studying what the child asks "why" about one can begin to see what kind of answers or solutions the child expects to receive.... A first general observation is that the child's whys bear witness to an intermediate precausality between the efficient cause and the final cause. Specifically, these questions seek reasons for phenomena which we see as fortuitous but which in the child arouse a need for a finalist explanation. "Why are there two Mount Salèves, a big one and a little one?" asked a six-year-old boy. To which almost all his contemporaries, when asked, gave the same answer: "the big one for big excursions and the small one for small" (Jean Piaget and B. Inhelder, *The Psychology of the Child*, trans. H. Weaver [New York: Basic Books, 1969], 109–10).

8. Compare chaps. 6–8 of the author's *Scepticism* (Oxford: Blackwell, 1980).

9. On the economic aspects of inquiry, see also the author's *Peirce's Philosophy of Science* (Notre Dame: University of Notre Dame Press, 1978), and in particular the last chapter, "Peirce and the Economy of Research."

10. W. Stanley Jevons, *The Principles of Science*, 2d ed. (London: Macmillan, 1874), 759.

11. Immanuel Kant, *Prolegomena to Any Future Metaphysic* (1783), sect. 57; Akad., 352. Italics in original.

12. Jevons, *Principles of Science*, 753.

13. On dialectical processes of this sort, see the author's *Dialectics* (Albany: State University of New York Press, 1977).

14. These considerations reemphasize that we must think of Q not as a *variable* (with a correlative range over "all questions") but as a *notational device* that acquires this status only in the context of a specification of a determinate range. Compare chapter 1 above.

15. William James, *The Will to Believe* (New York: Longmans, 1897), 22.

16. This chapter draws upon the author's *Empirical Inquiry* (Totowa, N.J.: Rowman and Littlefield, 1982).

Chapter 2

1. Karl R. Popper, *Objective Knowledge* (Oxford: Clarendon Press, 1944), 52–53.

2. Compare the discussion in Adolf Grünbaum, "Can a Theory Answer More Questions than One of Its Rivals?" *British Journal for the Philosophy of Science*, vol. 27 (1976), 1–22.

3. See Paul K. Feyerabend, *Against Method* (London: Humanities Press, 1975), 176.

4. Larry Laudan, "Two Dogmas of Methodology," *Philosophy of Science*, vol. 43 (1976), 585–97. See also his *Progress and Its Problems* (Berkeley, Los Angeles, London: University of California Press, 1978).

5. W. Stanley Jevons, *The Principles of Science*, 2d ed. (London: Macmillan, 1874), 754.

6. Quoted in Milic Capek, *The Philosophical Impact of Contemporary Physics* (Princeton, N.J.: Van Nostrand, 1961), xiv.

7. T. C. Mendenhall, *A Century of Electricity* (Boston and New York: Houghton Mifflin, 1887; 2d ed., revised, 1890), 223.

8. Max Planck, *Vortäge und Erinnerungen*, 5th ed. (Stuttgart: S. Hirzel, 1949), 169.

9. Thus, for Ostwald what "progress in discovery we experience anew from day to day ... affords a guarantee that in the course of time one query after another will be satisfied and one possibility after another will be realized, so that science will approach the ideal of omnipotence with rapid steps" (Wilhelm Ostwald, *Die Wissenschaft* [Leipzig, Akademische Verlagsgesellschaft, 1911], 47). Compare the following passage from a well-known textbook: "All branches of theoretical physics, with the exception of electricity and magnetism, can be regarded at the present state of science as concluded, that is, only immaterial changes occur in them from year to year." (Charles Emerson Curry, *Theory of Electricity and Magnetism* [New York and London: Macmillan, 1897], 1.) (I owe this reference to Martin Curd.)

10. J. J. Thomson, "Presidential Address to the British Association," *British Association for the Advancement of Science, Report for 1909*, 29.

11. J. J. Thomson, (1906), Michelson (1907), von Laue (1914), W. H. and W. L. Bragg (1915), Planck (1918), Einstein (1921), Bohr (1922), Heisenberg (1932), Dirac and Schrödinger (1933), Fermi (1938), Pauli (1945), Yukawa (1949).

12. Richard Feynman, *The Character of Physical Law* (Cambridge: Harvard University Press, 1965), 172–73.

13. Gunther S. Stent, *The Coming of the Golden Age* (Garden City, N.Y.: American Museum of Natural History, 1969), 111–13.

14. Compare Larry Laudan, *Progress and Its Problems* (Berkeley, Los Angeles, London: University of California Press, 1977).

15. The author's books, *Methodological Pragmatism* (New York: New York University Press, 1977), *Scientific Progress* (Oxford: Blackwell, 1978), and *Empirical Inquiry* (Totowa, N.J.: Rowman and Littlefield, 1982), provide further development and substantiation of this position.

Chapter 3

1. Larry Laudan, as cited in Ilkka Niiniluoto, "Scientific Progress," *Synthese*, vol. 45 (1980), 446. Rejection of the idea that science gets at the truth of things goes back to Karl Popper.

2. See the author's *Induction* (Oxford: Blackwell, 1980).

3. Michael E. Levin, "On Theory Change and Meaning Change," *Philosophy of Science*, vol. 46 (1979), 418.

4. This realization is something of which we can make no effective use: while we realize *that* many of our scientific beliefs are wrong, we have no way of telling *which* ones, and no way of telling *how* error has crept in.

5. Derek J. Price, *Science Since Babylon* (New Haven: Yale University Press, 1961), 137.

6. *The Structure of Scientific Revolutions*, 2d ed. (Chicago: University of Chicago Press, 1970). See also I. Lakatos and A. Musgrave, eds., *Criticism and the Growth of Knowledge* (Cambridge: Cambridge University Press, 1970).

7. The prime exponent of this position is Paul Feyerabend. See his essays, "Explanation, Reduction, and Empiricism," in Herbert Feigl and Grover Maxwell, eds., *Minnesota Studies in the Philosophy of Science*, vol. 3 (Minneapolis: University of Minnesota Press, 1962); "Problems of Empiricism," in R. G. Colodny, ed., *Beyond the Edge of Certainty* (Englewood Cliffs, N.J.: Prentice-Hall, 1965), 145–260; and "On the 'Meaning' of Scientific Terms," *Journal of Philosophy*, vol. 62 (1965), 266–74.

8. Practical problems have a tendency to remain structurally invariant. The sending of messages is just that, whether horse-carried letters or laser beams are used in transmitting the information.

9. This (essentially Baconian) idea that control over nature is the pivotal determinant of progress—in contrast to purely intellectual criteria (such as growing refinement, complication, or precision; let alone cumulation or proliferation)—has been mooted by several writers in response to Kuhn. See, for example, Paul M. Quay, "Progress as a Demarcation Criterion for Science," *Philosophy of Science*, vol. 41 (1974), 154–70 (especially 158); and also Friedrich Rapp, "Technological and Scientific Knowledge," in *Logic, Methodology, and Philosophy of Science: Proceedings in the 5th International Congress of DLMPS/IUHPS: London, Ontario, 1975* (Toronto: University of Toronto Press, 1976). The relevant issues are treated in depth in the present author's *Methodological Pragmatism* (Oxford: Blackwell, 1977).

10. For Hobbes's ideas in this region, see Hans Fiebig, *Erkenntnis und technische Erzeugung: Hobbes' operationale Philosophie der Wissenschaft* (Meisenheim am Glan: Zorn, 1973).

11. This final section draws on the author's *Scientific Progress* (Oxford: Blackwell, 1978).

Chapter 4

1. Herbert Spencer, *First Principles*, 7th ed. (London: Appleton's, 1889); see sects. 14–17 of part 2, "The Law of Evolution."

2. On the process in general see John H. Holland, *Hidden Order: How Adaptation Builds Complexity* (Reading: Addison Wesley, 1995). Regarding the specifically evolutionary aspect of the process see Robert N. Brandon, *Adaptation and Environment* (Princeton: Princeton University Press, 1990).

3. On the issues of this paragraph compare Stuart Kaufmann, *At Home in the Universe: To Search for the Laws of Self-Organization and Complexity* (New York and Oxford: Oxford University Press, 1995).

4. For further details see the author's *Induction* (Oxford: Blackwell, 1980).

5. An interesting illustration of the extent to which lessons in the school of better experience have accustomed us to expect complexity is provided by the contrast between the pairs: rudimentary/nuanced; unsophisticated/sophisticated; plain/elaborate; simple/intricate. Note that in each case the second, complexity-reflective alternative has a distinctly more positive (or less negative) connotation than its opposite counterpart.

6. Hans Reichenbach, *Experience and Prediction* (Chicago: University of Chicago Press, 1938), 376. Compare:

   Imagine that a physicist ... wants to draw a curve which passes through [points on a graph that represent] the data observed. It is well known that the physicist chooses the simplest curve; this is not to be regarded as a matter of convenience.... [For different] curves correspond as to the measurements observed, but they differ as to future measurements; hence they signify different predictions based on the same observational material. The choice of the simplest curve, consequently, depends on an inductive assumption: we believe that the simplest curve gives the best predictions.... If in such cases the question of simplicity plays a certain role for our decision, it is because we make the assumptions that the simplest theory furnishes the best predictions. (Ibid., 375–76)

7. *Science and Hypothesis* (New York: Dover Press, 1914), 145–46.

8. Note that in explaining the behavior of people we always presume normalcy and rationality on their part—a presumption that is, to be sure, defeasible and only holds "until proven otherwise."

9. Kant was the first philosopher clearly to perceive and emphasize this crucial point:

   But such a principle [of systematicity] does not prescribe any law for objects ..., it is merely a subjective law for the orderly management of the possessions of our understanding, that by the comparison of its concepts it may reduce them to the smallest possible number; it does not justify us in demanding from the objects such uniformity as will minister to the convenience and extension of our understanding; and we may not, therefore, ascribe to the [methodological or *regulative*] maxim ['Systematize knowledge!'] any objective [or descriptively *constitutive*] validity." (CPuR, A306 = B362.)

   Compare also C. S. Peirce's idea that the systematicity of nature is a regulative matter of scientific attitude rather than a constitutive matter of scientific fact. See Charles Sanders Peirce, *Collected Papers*, vol. 7 (Cambridge: Harvard University Press, 1958), sect. 7.134.

10. On this process see sections 2–3 of chapter 5.

11. See B. W. Petley, *The Fundamental Physical Constants and the Frontiers of Measurement* (Bristol: Hilger, 1985).

12. On the structure of dialectical reasoning see the author's *Dialectics* (Albany: State University of New York Press, 1977), and for the analogous role of such reasoning in philosophy see *The Strife of Systems* (Pittsburgh: University of Pittsburgh Press, 1985).

13. See John Dupré, *The Disorder of Things: Metaphysical Foundations of the Disunity of Science* (Cambridge: Harvard University Press, 1993).

14. See Steven Weinberg, *Dreams of a Final Theory* (New York: Pantheon, 1992). See also Edoardo Amaldi, "The Unity of Physics," *Physics Today*, vol. 261 (September 1973), 23–29. Compare also C. F. von Weizsäcker, "The Unity of Physics," in Ted Bastin, ed., *Quantum Theory and Beyond* (Cambridge: Cambridge University Press, 1971).

15. The older figures are given in S. S. Visher, "Starred Scientists, 1903–1943," in *American Men of Science* (Baltimore: Johns Hopkins, 1947). For many further details regarding the development of American science see the author's *Scientific Progress* (Oxford: Blackwell, 1978).

16. See, e.g., Derek J. Price, *Little Science, Big Science* (New York: Columbia University Press, 1963), 11.

17. Cf. Derek J. Price, *Science Since Babylon*, 2d ed. (New Haven: Yale University Press, 1975); see in particular chap. 5, "Diseases of Science."

18. Data from *An International Survey of Book Production During the Last Decades* (Paris: UNESCO, 1985).

19. Raymond Ewell, "The Role of Research in Economic Growth," *Chemical and Engineering News*, vol. 33 (1955), 2980–85.

20. It is worth noting for the sake of comparison that for more than a century now the *total* U.S. federal budget, its *nondefense subtotal*, and the aggregate budgets of all federal agencies concerned with the environmental sciences (Bureau of Mines, Weather Bureau, Army Map Service, etc.) have all grown at a uniform per annum rate of 9 percent. (See H. W. Menard, *Science: Growth and Change* [Cambridge: Harvard University Press, 1971], 188.)

21. Data from William George, *The Scientist in Action* (New York: Arno Press, 1938).

22. Du Pont's outlays for research stood at $1 million *per annum* during World War I (1915–1918), $6 million in 1930, $38 million in 1950, and $96 million in 1960. (Data from Fritz Machlup, *The Production and Distribution of Knowledge in the United States* [Princeton: Princeton University Press, 1962], 158–59, and see 159–60 for the relevant data on a larger scale.) In the United States at large, overall expenditures for scientific research and its technological development (R & D) stood at $.11 \times 10^9$ in 1920, and had risen to $.13 \times 10^9$ in 1930, $.38 \times 10^9$ in 1940, 2.9×10^9 in 1950, 5.1×10^9 in 1953–54, 10.0×10^9 in 1957–58, 11.1×10^9 in 1958–59, and had risen to ca. 14.0×10^9 in 1960–61 (ibid., 155 and 187). Machlup thinks it a not unreasonable conjecture that no other industry or economic activity in the United States has grown as fast as R & D (ibid., 155).

23. "Impact of Large-Scale Science on the United States," *Science*, vol. 134 (21 July 1961), 161–64 (see 161). Weinberg further writes: "The other main contender [apart from space exploration] for the position of Number One Event in the Scientific Olympics is high-energy physics. It, too, is wonderfully expensive (the Stanford linear accelerator is expected to cost $ 100×10^6), and we may expect to spend $ 400×10^6 per year on this area of research by 1970" (ibid., 164).

24. The "knowledge" at issue here need not be necessarily *correct*: it is merely *putative* knowledge that represents a comprehensively contrived best estimate of what the truth of the matter actually is.

25. Note that a self-concatenated concept or fact is still a concept or fact, even as a self-mixed color is still a color.

26. In information theory, *entropy* is the measure of the information conveyed by a message and is there measured by $k \log M$, where M is the number of structurally equivalent messages formulable with the available sorts of symbols. By analogy, the $\log I$ measure of the knowledge contained in a given body of information might be seen in the same light. Either way, the concept at issue measures informative actuality in relation to informative possibility. For there are two types of informative possibilities: (1) structural/syntactical as dealt with in classical information theory, and (2) hermeneutic/semantical (i.e., genuinely meaning oriented) as dealt with in the present discussion.

27. Some writers have suggested that the subcategory of significant information included in an overall body of crude data of size I should be measured by I^k (for some suitably adjusted value $0 < k < 1$)—for example by the "Rousseau's Law" standard of $\sqrt{I}$). (For details see chapter 6 of the author's *Scientific Progress* [Oxford: Blackwell, 1978].) Now since $\log I^k = k \log I$ which is proportional to $\log I$, the specification of *this* sort of quality level for information would again lead to a K a $\log I$ relationship.

28. See Stanislaw M. Ulam, *Adventures of a Mathematician* (New York: Scribner's, 1976).

29. Max Planck, *Vorträge und Erinnerungen*, 5th ed. (Stuttgart: S. Hirzel, 1949), 376; italics added. Shrewd insights seldom go unanticipated, so it is not surprising that other theorists should be able to contest claims to Planck's priority here. C. S. Peirce is particularly noteworthy in this connection.

30. It might be asked: "Why should a mere accretion in scientific 'information'—in mere belief—be taken to constitute *progress*, seeing that those later beliefs are not necessarily *true* (even as the earlier ones were not)?" The answer is that they are in any case better *substantiated*—that they are improvements on the earlier ones by way of the elimination of shortcomings. For a detailed consideration of the relevant issues, see the author's *Scientific Realism* (Dordrecht: D. Reidel, 1987).

31. The data here are set out more fully in the author's *Scientific Progress* (Oxford: Blackwell, 1978).

32. To be sure, we are caught up here in the usual cyclic pattern of all hypothetico-deductive reasoning. In addition to explaining the various phenomena we

have been canvassing, that projected K/I relationship is in turn substantiated by them. This is not a vicious circularity but simply a matter of the systemic coherence that lies at the basis of inductive reasonings.

33. See Henry Brooks Adams, *The Education of Henry Adams: An Autobiography* (Boston: Houghton Mifflin, 1918).
34. The situation with automobiles is analogous. Modern cars are simpler to operate (self-starting, self-shifting, power steering, etc.), but they are much more complex to manufacture, repair, maintain, and so on.
35. For other discussions relevant to this chapter's themes see also the author's *Scientific Progress* (Oxford: Blackwell, 1978) and *Cognitive Economy* (Pittsburgh: University of Pittsburgh Press, 1989).

Chapter 5

1. C. S. Peirce, *Collected Papers*, vol. 5, ed. C. Hartshorne and P. Weiss (Cambridge: Harvard University Press, 1934), sect. 7.144. See also Peirce's important 1898 paper on "Methods for Attaining Truth," in ibid., sects. 5.574 ff.
2. For Peirce's theory of the real truth as lying at the limit of our scientific findings, see the author's book *Peirce's Philosophy of Science* (Notre Dame and London: Notre Dame University Press, 1978). See also the interesting discussion in Ilkka Niiniluoto, "Scientific Progress," *Synthese* 45 (1980), 427–62.
3. For the historical background, see Laurens Laudan, "Peirce and the Trivialization of the Self-Correcting Thesis," in Ronald N. Giere and Richard S. Westfal, eds., *Foundations of Scientific Method: The 19th Century* (Bloomington: University of Indiana Press, 1973), 275–306.
4. It deserves emphasis, however, that this cookbook sort of a routine procedure for self-correctiveness was *not* operative in the ideas of C. S. Peirce.
5. Joseph Priestley, *The History and Present State of Electricity* (London: J. Dodsley, 1767), 381.
6. Georges Le Sage, "Quelques opuscules relatifs à la méthode," published posthumously in Pierre Provost, *Essais de philosophie*, vol. 2 (Paris: Alfand, 1804), 253–335.
7. This position also encounters immediately the difficulty of how this approximation-concept can be made to work in the case of theories or hypotheses. See note 24, below.
8. *Newton's Philosophy of Nature*, ed. H. S. Thayer (New York: Hafner Pub. Co., 1953), 117–18.
9. Arthur Millikan, *Science and the New Civilization* (New York: C. Scribner's Sons, 1936).
10. See, for example, George Sarton, *The Study of the History of Science* (Cambridge, Mass: Cambridge University Press, 1936), especially 5; and *History of Science and*

the New Humanism (Cambridge: Harvard University Press, 1937), especially 10–11. But William Wheweil was already an honorable exception to this rule. See his *Novum Organon Renovatum*, 3d ed. (London: J. W. Parker and Son, 1858), bk. 2, chap. 4, art. 9.

11. This list is taken from Larry Laudan, "A Confutation of Convergent Realism," *Philosophy of Science*, vol. 48 (1981), 38.

12. John N. Ziman, *Reliable Knowledge* (Cambridge: Cambridge University Press, 1978), 93–94.

13. This point is forcibly pressed by W. V. O. Quine (*Word and Object* [New York: Technology Press of the Massachusetts Institute of Technology, 1960], 23), who argues that talk of the limit of theories is based on an inappropriate mathematical analogy. (See note 24, below.)

14. See Paul Feyerabend, *Against Method* (London: Humanities Press, 1975); Thomas Kuhn, *The Structure of Scientific Revolutions* (Chicago: University of Chicago Press, 1962), and "Reflections on My Critics," in *Criticism and the Growth of Knowledge*, ed. I. Lakatos and A. Musgrave (Cambridge: Cambridge University Press, 1970), 91–195; and W. V. O. Quine, *Word and Object* (New York: Technology Press of the Massachusetts Institute of Technology, 1960).

15. Karl Popper, "Truth, Rationality and the Growth of Scientific Knowledge," in *Conjectures and Refutations* (New York: Basic Books, 1962), 125.

16. Larry Laudan, *Progress and Its Problems* (Berkeley, Los Angeles, London: University of California Press, 1977). Laudan is one of the most eloquent and cogent critics of the convergentism that our present discussion also rejects. See also his paper "A Confutation of Convergent Realism," *Philosophy of Science*, vol. 48 (1981), 19–49.

17. This shibboleth of the contemporary philosophy of science is not all that new. Already at the turn of the century, Sir Michael Foster wrote:

> The path [of progress in science] may not be always a straight line; there may be swerving to this side and to that; ideas may seem to return again and again to the same point of the intellectual compass; but it will always be found that they have reached a higher level—they have moved, not in a circle, but in a spiral. Moreover, science is not fashioned as is a house, by putting brick to brick, that which is once put remaining as it was put to the end. The growth of science is that of a living being. As in the embryo, phase follows phase, and each member or body puts on in succession different appearances, though all the while the same member, so a scientific conception of one age seems to differ from that of a following age. ("The Growth of Science in the Nineteenth Century," *Annual Report of the Smithsonian Institution for 1899* [Washington, D.C., 1901], 175 [as reprinted from Foster's 1899 presidential address to the British Association for the Advancement of Science])

18. Anonymous, "The Nature of Knowledge," *The Economist*, no. 281 (Dec. 26, 1981–Jan. 8, 1982), 103.

19. Ernan McMullin. "Limits of Scientific Inquiry," in J. C. Steinhardt, ed., *Science and the Modern World* (New York: Plenum Press, 1966), 68.

20. See Larry Laudan, "Two Dogmas of Methodology," *Philosophy of Science*, vol. 43 (1976), 585–97, and *Progress and Its Problems* (Berkeley, Los Angeles, London: University of California Press, 1977).

21. Bentley Glass, "Science: Endless Horizons or Golden Age?" *Science*, vol. 171 (1971), 25–29.

22. Ibid., 24.

23. The geographic-exploration analogy is an old standby: "[Science] cannot keep on going so that we are always going to discover more and more new laws.... It is like the discovery of America—you only discover it once. The age in which we live is the age in which we are discovering the fundamental laws of nature, and that day will never come again." (Richard Feynman, *The Character of Physical Law* [Cambridge: Harvard University Press, 1965], 172). See also Gunther Stent, *The Coming of the Golden Age* (Garden City, N.Y.: Natural History Press, 1969), and S. W. Hawking, "Is the End in Sight for Theoretical Physics?" *Physics Bulletin*, vol. 52 (1981), 15–17.

24. Marxist theoreticians take this view very literally—in the manner of Lenin's idea of the "inexhaustibility" of matter in *Materialism and Empirico-Criticism*. Purporting to inherit from Spinoza a thesis of the infinity of nature, they construe this to mean that any cosmology that denies that infinite spatial extension of the universe must be wrong.

25. "Remarks by D. Bohm," in *Observation and Interpretation*, ed. S. Körner (New York and London: Routledge, 1957), 56. For a fuller development of Bohm's views on the "qualitative infinity of nature," see his *Causality and Chance in Modern Physics* (New York: Van Nostrand, 1957).

26. A later writer put it this way: "Go on as far as we will, in the subdivision of continuous quantity, yet we never get down to the absolute point. Thus scientific method leads us to the inevitable conception of an infinite series of successive orders of infinitely small quantities. If so, there is nothing impossible in the existence of a myriad universes within the compass of a needle's point, each with its stellar systems, and its suns and planets, in number and variety unlimited" (W. S. Jevons, *The Principles of Science*, 2d ed. [London: Macmillan, 1874], 767).

27. The idea of a succession of "layers of depth" in the analysis of nature, each giving rise to its own characteristic body of laws in such a way that each of these law-manifolds is more encompassing than its predecessors and that their successive discovery represents a sequentially deeper penetration of the structure of nature was, so far as I know, first mooted in contemporary physics in E. P. Wigner, "The Limits of Science," *Proceedings of the American Philosophical Society*, vol. 94 (1950), 422–27.

28. Sir Denys Wilkinson, *The Quarks and Captain Ahab or: The Universe as Artifact* (Palo Alto: Stanford University Press, 1977; Schiff Memorial Lecture), 4–5.

29. The French physicist Jean-Paul Vigier, for example, has revived this idea of Pascal's along Hegelian lines, arguing as follows: "We would prefer to say that at all levels of

Nature you have a mixture of causal and statistical laws (which come from deeper or external processes). As you progress from one level to another you get new qualitative laws. Causal laws at one level can result from averages of statistical behavior at a deeper level, which in turn can be explained by deeper causal behavior, and so on ad infinitum. If you then admit that Nature is infinitely complex and that in consequence no final stage of knowledge can be reached, you see that at any stage of scientific knowledge causal and probability laws are necessary to describe the behavior of any phenomenon, and that any phenomenon is a combination of causal and random properties inextricably woven with one another. All things in Nature then appear as a dialectical synthesis of the infinitely complex motions of matter out of which they surge and grow and into which they finally are bound to disappear" ("The Concept of Probability in the Frame of the Probabilistic and the Causal Interpretation of Quantum Mechanics," in *Observation and Interpretation*, ed. S. Körner [New York and London: Routledge, 1957], 77). The idea of a succession of "layers of depth in the analysis of nature, each giving rise to its own characteristic body of laws in such a way that each of these law-manifolds is more encompassing than its predecessors and that their successive discovery represents a sequentially deeper penetration of the structure of nature, is not peculiar to Vigier. It was, so far as I know, first mooted in contemporary physics in E. P. Wigner, "The Limits of Science," *Proceedings of the American Philosophical Society*, vol. 94 (1950), 422–27.

30. "If by the 'infinite complexity of nature' is meant only the infinite multiplicity of the *phenomena* it contains, there is no bar to final success in theory making, since theories are not concerned with particulars as such. So too, if what is meant is only the infinite variety of natural phenomena, . . . that too may be comprehended in a unitary theory" ("Scientific Revolutions for Ever?" *British Journal for the Philosophy of Science* 19 [1967]: 41). For an interesting and suggestive analysis of "the architecture of complexity," see Herbert A. Simon, *The Sciences of the Artificial* (Cambridge: MIT Press, 1969).

31. Recall Goethe's stricture:

> Natur hat weder Kern noch Schale, Alles ist sie mit einem Male.
>
> —*Geflügelte Worte*

32. To be sure, nature-as-it-is cannot, in the final analysis, be simpler than nature-as-it-is-thought-to-be, because the latter is, in its fashion, part of the former.

33. John Herschel, *Familiar Lectures on Scientific Subjects* (London: A. Strahan, 1867), 458.

34. The idea that our knowledge about the world reflects an *interactive* process, to which both the object of knowledge (the world) and the knowing *subject* (the inquiring mind) make essential and ultimately inseparable contributions, is elaborated in the author's *Conceptual Idealism* (Oxford: Blackwell, 1975).

35. W. Stanley Jevons, *The Principles of Science*, 2d ed. (London: Macmillan, 1874), 755.

36. Compare D. A. Bromley's observation:

> Even if physicists could be sure that they had identified all the particles that can exist, some

obviously fundamental questions would remain. Why, for instance, does a certain universal ratio in atomic physics have the particular value 137.036 and not some other value? This is an experimental result: the precision of the experiments extends today to these six figures. Among other things, this number relates the extent or size of the electron to the size of the atom, and that in turn to the wavelength of light emitted. From astronomical observation it is known that this fundamental ratio has the same numerical value for atoms a billion years away in space and time. As yet there is no reason to doubt that other fundamental ratios, such as the ratio of the mass of the proton to that of the electron, are as uniform throughout the universe as is the geometrical ratio $P = 3.14159$. Could it be that such physical ratios are really, like P, mathematical aspects of some underlying logical structure? If so, physicists are not much better off than people who must resort to wrapping a string around a cylinder to determine the value of P! For theoretical physics thus far sheds hardly a glimmer of light on this question. (D. A. Bromley et al., *Physics in Perspective*, Student Edition [Washington, D.C.: National Academy of Sciences/National Research Council, 1975], 28)

37. E. P. Wigner, "The Limits of Science," *Proceedings of The American Philosophical Society*, vol. 94 (1950), 422–27.

38. Quoted in Ilkka Niiniluoto, "Scientific Progress," *Synthese*, vol. 45 (1980), 435.

39. In this regard, the position of E. P. Wigner seems altogether right-minded:

    in order to understand a growing body of phenomena, it will be necessary to introduce deeper and deeper concepts into physics and that this development will not end by the discovery of the final and perfect concepts. I believe that this is true: we have no right to expect that our intellect can formulate perfect concepts for the full understanding of inanimate nature's phenomena. ("The Limits of Science," *Proceedings of the American Philosophical Society*, vol. 94 (1950), 424)

40. The ideas of this paragraph are developed at greater length in the author's *Scientific Progress* (Oxford: Blackwell, 1978).

41. W. Stanley Jevons, *The Principles of Science*, 2d ed. (London: Macmillan, 1874), 753–54.

42. The present critique of convergentism is thus very different from that of W. V. O. Quine. He argues that the idea of "convergence to a limit" is defined for numbers but not for theories, so that speaking of scientific change as issuing in a "convergence to a limit" is a misleading metaphor. "There is a faulty use of a mathematical analogy in speaking of a limit of theories, since the notion of a limit depends on that of a 'nearer than,' which is defined for numbers and not for theories" (*Word and Object* [New York: Technology Press of the Massachusetts Institute of Technology, 1960], 23). I myself am perfectly willing to apply the metaphor of substantial and insignificant differences to theories but am simply concerned to deny, as a matter of fact, that the course of scientific theory-innovation must eventually descend to the level of trivialities.

Chapter 6

1. The example is adapted from John L. Casti, *Searching for Certainty* (New York: W. Morris and Co., 1990), 42.
2. For the halting problem, see C. A. R. Hoare and D. C. S. Allison, "Incomputability," *Computing Surveys*, vol. 4, no. 3 (September 1972).
3. This point has been argued forcefully in other contexts by K. R. Popper. See his *The Poverty of Historicism* (London: Routledge and Kegan Paul, 1957), vi. But also compare Peter Urbach, "Is Any of Popper's Arguments Against Historicism Valid?" *British Journal for the Philosophy of Science*, vol. 29 (1978), 117–30 (see 127–28). Popper's argument has its problems, however. It pivots on Kurt Gödel's famous Incompleteness Theorem to the effect that no reasoner proceeding on mathematical principles can decide whether a certain sentence G is or is not a theorem of arithmetic. Popper now proposes asking the putative predictor the question: "If you are asked at t the question 'Is G a theorem?' (and we attend until t' for an answer) will you or will you not before t' respond with 'yes' or 'no'?" (See also Popper's "Indeterminism in Classical Physics and in Quantum Physics," *British Journal for the Philosophy of Science*, vol. 1 (1950), 117–33 and 173–95 (see 183). The problem with this particular example is that the temporality at issue is spurious. For the predictor's incapacity to answer the given question—other than by "can't say"—roots in its incapacity to answer its detemporalized counterpart: "If asked 'Is G a theorem?' will you answer 'yes' or 'no'?" Given the speciousness of its temporalization, the question at issue is not genuinely predictive. Its incapacity to answer may render that prediction cognitively imperfect, but it is not predictively imperfect.
4. On this issue see chapter 6, above.

Chapter 7

1. See J. Bernstein, *Three Degrees Above Zero* (New York: Scribner, 1984).
2. Frederick Soddy, *Science and Life* (New York: E. P. Dutton, 1930).
3. Quoted in *Daedalus*, vol. 107 (1978), 24.
4. As one commentator has wisely written: "But prediction in the field of pure science is another matter. The scientist sets forth over an uncharted sea and the scribe, left behind on the dock, is asked what he may find at the other side of the waters. If the scribe knew, the scientist would not have to make his voyage" (Anonymous, "The Future as Suggested by Developments of the Past Seventy-Five Years," *Scientific American*, vol. 123 [1920], 321). The role of unforeseeable innovations in science forms a key part of Popper's case against the unpredictability of man's social affairs—given that new science engenders new technologies, which in turn make for new modes of social organization. (See K. R. Popper, *The Poverty of Historicism* [London: Routledge and Kegan Paul, 1957], vi and passim.) The impredictability of revolutionary changes in science also figures centrally in W. B. Gallie's "The Lim-

its of Prediction," in S. Körner, ed., *Observation and Interpretation* (New York: Academic Press, 1957). Gallie's argumentation is weakened, however, by a failure to distinguish between the generic fact of future discovery in a certain domain and its specific nature. See also Peter Urbach, "Is Any of Popper's Arguments Against Historicism Valid?" *British Journal for the Philosophy of Science*, vol. 29 (1978), 117–30 (see 128–29), whose deliberations seem (to this writer) to skirt the key issues. A judicious and sympathetic treatment is given in Alex Rosenberg, "Scientific Innovation and the Limits of Social Scientific Prediction," *Synthese*, vol. 97 (1993), 161–81. On the present issue Rosenberg cites the instructive anecdote of the musician who answered the question "Where is jazz heading?" with the response: "If I knew that, I'd be there already" (op. cit., 167).

5. Quoted in Philip Handler, ed., *Biology and the Future of Man* (Oxford: Clarendon Press, 1970), 165.

6. This holds for technology as well through the principle of "equivalent invention" described by S. C. Gilfillan in 1939. (For the references, see S. Colum Gilfillan, "A Sociologist Looks at Technical Prediction," in James R. Bright, ed., *Technical Forecasting for Industry and Government* [Englewood Cliffs, N.J.: Prentice Hall, 1968], 3–34). Instancing the problem of flying aircraft in fog, Gilfillin notes that some dozen different ways of addressing it were under consideration in the 1930s. But which method would prove most successful and thus prevail was imponderable. That the problem would be resolved could be predicted with confidence, but *how* was unforeseeable.

7. For further aspects regarding this chapter's issues see the author's *Predicting the Future* (Albany: State University of New York Press, 1997).

8. Baden Powell, *Essays on the Spirit of the Inductive Philosophy* (London: Longman, 1855), 23.

9. A. N. Whitehead, as cited in John Ziman, *Reliable Knowledge* (Cambridge: Cambridge University Press, 1969), 142–43.

10. See Thomas Kuhn, *The Structure of Scientific Revolutions*, 2d ed. (Chicago: University of Chicago Press, 1970), for an interesting development of the normal/revolutionary distinction.

11. See also the criticisms of his argument in David Bohm, *Causality and Chance in Modern Physics* (London: Routledge, 1957), 95–96.

12. Larry Laudan, *Progress and Its Problems* (Berkeley, Los Angeles, London: University of California Press, 1977), 59–60.

13. Cicero, *De finibus*, I, vi. 19.

14. See R. G. Collingwood, *An Essay in Metaphysics* (Oxford: Clarendon Press, 1940), and *The Idea of Nature* (Oxford: Clarendon Press, 1945).

15. *Über die Grenzen des Naturekennens*, 11th ed. (Leipzig: Veit and Co., 1916).

16. Powell, *Essays*, 111.

Chapter 8

1. Or perhaps alternatively: always after a certain time—at every stage subsequent to a certain juncture.
2. In terms of the symbolism introduced in chapter 2, this thesis comes to:
   $(\forall t)(\exists Q)[Q \in \mathbf{Q}(S_t) \ \& \sim S_t @ Q]$.
3. In terms of the symbolism introduced in chapter 2, this thesis comes to:
   $(\exists Q)(\forall t)[Q \in \mathbf{Q}(S_t) \ \& \sim S_t @ Q]$.
4. In terms of the symbolism of chapter 2, this thesis claims the existence of one specifiable question Q of such a sort that:
   $(\forall t) \sim S_t @ Q$.
5. This work was published together with a famous prior (1872) lecture on the limits of scientific knowledge as *Über die Grenzen des Naturekennens: Die Sieben Welträtsel—Zwei Vorträge*, 11th ed. (Leipzig: Veit and Co., 1916). The earlier lecture has appeared in English translation as "The Limits of Our Knowledge of Nature," *Popular Science Monthly*, vol. 5 (1874), 17–32. For du Bois-Reymond, see Ernst Cassirer, *Determinism and Indeterminism in Modern Physics: Historical and Systematic Studies of the Problem of Causality* (New Haven: Yale University Press, 1956), part 1.
6. Bonn, 1889, trans. by J. McCabe as *The Riddle of the Universe—at the Close of the Nineteenth Century* (New York and London: Harper and Bros., 1901). On Haeckel, see the article by Rollo Handy in *The Encyclopedia of Philosophy*, ed. Paul Edwards, vol. 3 (New York: Macmillan, 1967).
7. Haeckel, *Die Welträtsel*, 365–66.
8. *The Grammar of Science* (London: A. and C. Black, 1892), sect. 7.
9. Carl G. Hempel, "Science Unlimited," *Annals of the Japan Association for Philosophy of Science*, vol. 14 (1973), 200 (italics added).
10. Note, too, that the question of the existence of facts is a horse of a very different color from that of the existence of things. There being no things is undoubtedly a possible situation; there being no *facts* is not (since if the situation were realized, this would itself constitute a fact).
11. *Dialogues Concerning Natural Religion*, ed. N. K. Smith (London, 1920), 189.
12. G. W. Leibniz, "Principles de la nature et de la grace," sect. 8 (italics added).
13. For criticisms of ways of avoiding the question, "Why is there something rather than nothing?" see chapter 3 of William Rowe, *The Cosmological Argument* (Princeton: Princeton University Press, 1975).
14. This has been argued in detail by the Canadian philosopher John Leslie. See his papers: "The World's Necessary Existence," *International Journal for the Philosophy of Religion*, vol. 11 (1980), 297–329; "Efforts to Explain All Existence," *Mind*, vol. 87

(1978), 181–97; "The Theory that the World Exists Because It Should," *American Philosophical Quarterly*, vol. 7 (1910), 286–98; and "Anthropic Principle, World Ensemble, Design," *American Philosophical Quarterly*, vol. 19 (1982), 141–51; as well as his book, *Value and Existence* (Totowa, N.J.: Rowman and Littlefield, 1979).

15. For an interesting attempt to deal with the question, see chapter 2 of Robert Nozick, *Philosophical Explanations* (Cambridge: Harvard University Press, 1981).

16. W. Stanley Jevons, *The Principles of Science*, 2d ed. (London: Macmillan, 1874), 764.

17. This issue is the second of Reymond's seven "riddles of the universe."

18. For the anthropic principle, see George Gale, "The Anthropic Argument," *Scientific American*, vol. 245, December 1981, 154–71. See also the literature cited in Leslie's papers, cited in note 14, above.

19. Charles Sanders Peirce, *Collected Papers*, ed. C. Hartshorne et al., vol. 6 (Cambridge: Harvard University Press, 1929), sect. 6.556.

20. Baden Powell, *Essays on the Spirit of the Inductive Philosophy* (London: Longman, 1855) 106–107.

Chapter 9*

1. See Steven Weinberg, *Dreams of a Final Theory* (New York: Pantheon, 1992). See also Edoardo Amaldi, "The Unity of Physics," *Physics Today*, vol. 261 (September 1973), 23–29. Compare also C. F. von Weizsäcker, "The Unity of Physics" in Ted Bastin, ed., *Quantum Theory and Beyond* (Cambridge: Cambridge University Press, 1971).

2. Op. cit., 212.

3. It is questionable, of course, whether it makes sense to contemplate quantifying over truths if we open up the range of facts to the entire spectrum of mathematical potentiality. However, if we limit our purview to the more limited range of facts about the natural world it would seem to become a practicable proposition.

4. The model was initially promulgated in Carl G. Hempel and Paul Oppenheim, "Studies in Logic of Explanation," *Philosophy of Science*, vol. 15 (1948), 135–75. The historical background of this paper is set out in the author's "H_2O: Hempel-Helmer-Oppenheim: An Episode in the History of Scientific Philosophy in the Twentieth Century" *Philosophy of Science*, vol. 15 (1997), 779–805; reprinted in his *Profitable Speculations* (Lanham, Md.: Rowman and Littlefield, 1997), 69–107. The subsequent development of thought about explanation in the middle part of the twentieth century is vividly depicted in Wesley Salmon's *From Decades of Scientific Explanation* (Minneapolis: University of Minnesota Press, 1990).

5. Hempel-Oppenheim originally said "deductively" where we have "inferentially." Hempel later loosened this to contemplate the proposal of probabilistic explanations.

6. On this principle in its relation to the cosmological aspect for the existence of God, see William L. Rowe, *The Cosmological Argument* (Princeton: Princeton University

Press, 1975). See also Richard M. Gale, *On the Nature and Existence of God* (Cambridge: Cambridge University Press, 1991); and Alexander R. Pruss, "The Hume-Edwards Principle and the Cosmological Argument," *International Journal for Philosophy of Religion*, vol. 434 (1988), 149–65.

7. In a way, the crux is whether that set is defined extensionally (by way of an inventory) or intensionally (as the set of all items that have the property φ). In the former case the Hume-Edwards principle holds; in the latter it does not. The existence of each of the things that have φ may be established without ever mentioning φ. But if we ask for an explanation of the existence of the set of φ-possessors then what we want *(inter alia)* is an answer to the generic question of why φ-possessors exist as such. And of course in asking for an explanation of the existence of "this particular world," "this world's things," or "this worlds events/states" (that is, of the world's spatiotemporal constituents) we are dealing with something that cannot be inventoried (extensionally specified).

8. It might seem plausible to "weaken" [C] to read:

   Whenever a fact is explicable at all—i.e., is not simply a "basic fact of nature"—then T* affords its explanation:

   $(\forall t)[(\exists t') \ t' \Sigma t \to T^* \Sigma t\ ]$

   However in the presence of the PSR to the effect that $(\forall t)(\exists t) \ t' \Sigma t$ this simply comes down to [C] itself. Once we adopt PSR we abjure any prospect of "fundamental" (i.e., themselves inexplicable) facts of nature.

9. It is sometimes suggested that the Anthropic Hypothesis (AH) qualifies as an ultimate theory. However, this is not the case. The AH does not explain the laws of nature but rather only seeks to account for the initial conditions for cosmic escalation on the basis of arguing that given that the fundamental laws of nature are as we suppose them to be certain initial conditions must be postulated if those laws (as supposed) are to account for the evaluation of intelligent beings in the cosmos. For from *explaining* the laws of nature this line of reasoning simply *presupposes* them—what it explains (of most and not best) are the initial conditions of cosmic evolution. (On the Anthropic Hypothesis see John D. Barrow and Frank T. Tipler, *The Anthropic Cosmological Principle* [Oxford: Clarendon Press, 1992].)

10. See T. S. Champlain, *Reflexive Paradoxes* (London: Routledge, 1988), and S. J. Bartlett and P. Suber, eds., *Self-Reference* (Dordrecht: Nijhoff, 1987). See also Thomas Breuer, "Universal und unvollständig: Theorien über alles?" *Philosophia Natrualis*, vol. 34 (1997), 1–20.

11. K. R. Popper, *Objective Knowledge: An Evolutionary Approach* (Oxford: Clarendon Press, 1972), 195.

12. Compare the author's *Cognitive Systematization* (Oxford: Blackwell, 1980) for a fuller development of these ideas.

13. An analogy may help: that of the poet who in writing a poem to formulate his principles of prosody also manages to present a paradigmatic illustration of them.

14. Compare the author's *Methodological Pragmatism* (Oxford: Blackwell, 1973).
15. A discussion that reaches much the same doctrinal destination by a very different route—one that proceeds via predominantly historical rather than logico-theoretical considerations—is Philip Kitcher's "Explanatory Unification," *Philosophy of Science*, vol. 48 (1981), 507–31.

Chapter 10

1. The author's *Cognitive Systematization* (Oxford: Blackwell, 1979) deals with these matters.
2. One possible misunderstanding must be blocked at this point. To learn about nature, we must interact with it. And so, to determine some feature of an object, we may have to make some impact upon it that would perturb its otherwise obtaining condition. (The indeterminacy principle of quantum theory affords a well-known reminder of this.) It should be clear that this matter of physical interaction for data acquisition is not contested in the ontological indifference thesis here at issue.
3. Note that this is independent of the question "Would we ever want to do so?" Do we ever want to answer all those predictive questions about ourselves and our environment, or are we more comfortable in the condition in which "ignorance is bliss"?
4. S. W. Hawking, "Is the End in Sight for Theoretical Physics?" *Physics Bulletin*, vol. 32 (1981), 15–17.
5. This sentiment was abroad among physicists of the *fin de siècle* era of 1890–1900. (See Lawrence Badash, "The Completeness of Nineteenth-Century Science," *Isis*, vol. 63 [1972], 48–58.) Such sentiments are coming back into fashion today. See Richard Feynman, *The Character of Physical Law* (Cambridge: MIT Press, 1965), 172. See also Gunther Stent, *The Coming of the Golden Age* (Garden City, N.Y.: Natural History Press, 1969); and S. W. Hawking, "Is the End in Sight for Theoretical Physics?" *Physics Bulletin*, vol. 32, 1981, 15–17.
6. See Eber Jeffrey, "Nothing Left to Invent," *Journal of the Patent Office Society*, vol. 22 (July 1940), 479–81.
7. For this inference could only be made if we could move from a thesis of the format $\sim(\exists r)(r \in S \,\&\, r \Rightarrow p)$ to one of the format $(\exists r)(r \in S \,\&\, r \Rightarrow \sim p)$, where "$\Rightarrow$" represents a grounding relationship of "furnishing a good reason" and p is, in this case, the particular thesis "S will at some point require drastic revision." That is, the inference would go through only if the lack (in S) of a good reason for p were itself to assure the existence (in S) of a good reason for $\sim p$. But the transition to this conclusion from the given premise would go through only if the former, antecedent fact itself constituted such a good reason; that is, only if we had $\sim(\exists r)(r \in S \,\&\, r \Rightarrow p) \Rightarrow \sim p$. Thus, the inference would go through only if, by contraposition, $p \Rightarrow (\exists r)(r \in S \,\&\, r \Rightarrow p)$. This thesis claims that the very truth of p will itself be a good reason to hold that S affords a good reason for p—in sum, that S is probatively complete with regard to truths in general.

8. Immanuel Kant, *Critique of Practical Reason*, 122 [Akad.].

9. See the author's *Peirce's Philosophy of Science* (Notre Dame: University of Notre Dame Press, 1978).

10. Note, however, that to say that some ideal can be legitimated by practical considerations is not to say that all ideals must be legitimated in this way.

11. For some recent discussions of scientific realism, see Wilfred Sellars, *Science Perception and Reality* (London: Humanities Press, 1963); E. McKinnon, ed., *The Problem of Scientific Realism* (New York: Appleton-Century-Crofts, 1972); Rom Harré, *Principles of Scientific Thinking* (Chicago: University of Chicago Press, 1970); and Frederick Suppe, ed., *The Structure of Scientific Theories*, 2d ed. (Urbana: University of Illinois Press, 1977).

12. Some of the issues of this discussion are developed at greater length in the author's *Methodological Pragmatism* (Oxford: Blackwell, 1977), *Scientific Progress* (Oxford: Blackwell, 1979), and *Cognitive Systematization* (Oxford: Blackwell, 1979).

13. Keith Lehrer, "Review of *Science, Perception, and Reality* by Wilfred Sellars," *The Journal of Philosophy*, vol. 63 (1966), 269.

14. This chapter draws on the author's *Empirical Inquiry* (Totowa, N.J.: Rowman and Littlefield, 1982).

Chapter 11

1. Sir Denys H. Wilkinson, *The Quarks and Captain Ahab or: The Universe as Artifact* (Palo Alto: Stanford University Press, 1977; Schiff Memorial Lecture), 12–13.

2. *Collected Papers*, ed. C. Hartshorne et al., vol. 1 (Cambridge: Harvard University Press, 1931), 44–45 (sects. 108–109); ca. 1896.

3. "Looking back, one has the impression that the historical development of the physical description of the world consists of a succession of layers of knowledge of increasing generality and greater depth. Each layer has a well defined field of validity; one has to pass beyond the limits of each to get to the next one, which will be characterized by more general and more encompassing laws and by discoveries constituting a deeper penetration into the structure of the Universe than the layers recognized before." (Edoardo Amaldi, "The Unity of Physics," *Physics Today*, vol. 261, no. 9 [September 1973], 24.) See also E. P. Wigner, "The Unreasonable Effectiveness of Mathematics in the Natural Sciences," *Communications on Pure and Applied Mathematics*, vol. 13 (1960), 1–14, as well as his "The Limits of Science," *Proceedings of the American Philosophical Society*, vol. 93 (1949), 521–26. Compare also chapter 8 of Henry Margenau, *The Nature of Physical Reality* (New York: McGraw-Hill, 1950).

4. D. A. Bromley et al., *Physics in Perspective,* Student Edition (Washington, D.C.: National Research Council/National Academy of Science Publications, 1973), 23.

5. Ibid.

6. Gerald Holton, "Models for Understanding the Growth and Excellence of Scientific Research," in Stephen R. Graubard and Gerald Holton, eds., *Excellence and Leadership in a Democracy* (New York: Columbia University Press, 1962), 115.
7. See Derek J. deSolla Price, *Little Science, Big Science* (New York: Columbia University Press, 1963), 92–93 and 101ff.
8. Compare Keith Norris and John Vaizey, *The Economics of Research and Technology* (London: Allen and Unwin, 1973), 164.
9. This latter relationship has been verified for many industries, including the production of fuels, chemicals, and metals. See, for example, the data cited in M. Korach, "The Science of Industry," in *The Science of Science*, ed. M. Goldsmith and A. Mackey (London, 1964), 179–94.
10. Copenhagen and Madison: Lubrecht and Cramer, Limited, 1963.
11. W. Stanley Jevons, *The Principles of Science*, 2d ed. (London: Macmillan, 1874), 752–53.
12. The present chapter draws on chapter 10 of the author's *Scientific Progress* (Oxford: Blackwell, 1978). A much fuller treatment of practical limitations to science is given in that book.

Chapter 12

1. The salient point is that unless I can *tell* (i.e., myself be able to claim justifiedly) that you are justified in your claim, I have no adequate grounds to accept it: it has not been made credible to me, irrespective of how justified you may be in regard to it. To be sure, for your claim to be credible for me I need not know *what* your justification for it is, but I must be in a position to realize *that* you are justified. Your reasons may even be incomprehensible to me, but for credibility I require a rationally warranted assurance—perhaps only on the basis of general principles—that those reasons are both extant and cogent.
2. On unsolvable calculating problems, mathematical completeness, and computability, see Martin Davis, *Computability and Unsolvability* (New York: McGraw-Hill, 1958; expanded reprint edition, New York: Dover, 1982). See also N. B. Pour-El and J. I. Richards, *Computability in Analysis and Physics* (Berlin: Springer Verlag, 1989) or on a more popular level, Douglas Hofstadter, *Gödel, Escher, Bach: An Eternal Golden Braid* (New York: Basic Books, 1979).
3. Some problems are not inherently unsolvable but cannot in principle be settled by computers. An instance is "What is an example of a word that no computer will ever use?" Such problems are inherently computer-inappropriate and for this reason a failure to handle them satisfactorily also cannot be seen as a meaningful limitation of computers.
4. On Gödel's theorem see S. G. Shanker, ed., *Gödel's Theorems in Focus* (London: Croom Helm, 1988), a collection of essays that provide instructive, personal, philosophical, and mathematical perspectives on Gödel's work.

5. On this issue see Hans J. Brennerman, "Complexity and Transcomputability" in Ronald Duncan and Miranda Weston-Smith, eds., *The Encyclopedia of Ignorance* (Oxford: Pergamon Press, 1977), 167–74.
6. For a comprehensive survey of the physical limitations of computers see Theodore Leiber, "Chaos, Berechnungskomplexität und Physik: Neue Grenzen wissenschaftlicher Erkenntnis," *Philosophia Naturalis*, vol. 34 (1997), 23–54.
7. As stated, this question involves a bit of anthropomorphism in its use of "you." But this is so only for reasons of stylistic vivacity. That "you" is, of course, only shorthand for "computer number such-and-such."
8. On the inherent limitation of predictions see the author's *Predicting the Future* (Albany: State University of New York Press, 1997).
9. Note that T_1 is weaker than:

    T_3: *No computer can reliably determine that there are substantive problems that are computer irresolvable:*

    $\sim(\exists C)\ C\ \det\ (\exists P)(\forall C)\ \sim C\ \mathrm{res}\ P$

    This stronger thesis is surely false, but the truth of the weaker T_1 nevertheless remains intact.
10. See Solomon Feferman et al., eds., Kurt Gödel, *Collected Works*, vol. 3: *Unpublished Essays and Lectures* (Oxford: Oxford University Press, 1995), and especially the 1951 paper on "Some Basic Theorems on the Foundation of Mathematics and Their Indications."
11. On Church's thesis see Hartley Rogers, *Theory of Recursive Functions and Effective Computability* (New York: McGraw-Hill, 1967), and Martin Davis, ed., *The Undecidable* (New York: Raven Press, 1965).
12. On Turing machines see R. Hesken, ed., *The Universal Turing Machine* (Oxford: Oxford University Press, 1988).
13. But could one not simply connect computers up with one another so as to create one vast megacomputer that could thereby do anything that any computer can? Clearly this can (in theory) be done when the set of computers at issue includes "all currently existent computers." But of course we cannot throw future ones into the deal, let alone merely possible ones.
14. Cambridge: MIT Press, 1992. This book is an updated revision of his earlier *What Computers Can't Do* (New York: Harper Collins, 1972).
15. For some discussion of this issue from a very different point of approach see Roger Penrose, *The Emperor's New Mind* (New York: Oxford University Press, 1989).
16. We have just claimed P_5 as computer-irresolvable. This contention, of course, entails $(\exists P)(\forall C)\sim C\ \mathrm{res}\ P$ or equivalently $\sim(\forall P)(\exists C)C\ \mathrm{res}\ P$. Letting this thesis be T_3, we may recall that T_2 comes to $(\forall C)\sim C\ \det\ \sim T_3$. If T_3 is indeed true, then this contention—that is, T_2—will of course immediately follow.
17. Someone might suggest: "But can one use the same line of thought to show that there are computer-solvable problems that people cannot possibly solve by simply

interchanging the reference to 'computers' and 'people' throughout its preceding argumentation?" But this will not do. For the fact that it is people that use computers means that one can credit people with computer-provided problem solutions via the idea that *people can solve problems with computers*. The reverse cannot be claimed with any semblance of plausibility. The situation is not in fact symmetrical and so the proposed interchange will not work. This issue is elaborated in the next section.

18. Such examples were to come along only later with the relating of Gödel's results to number-theoretic issues relating to the solution of Diophantine equations. For a good expository account see Martin D. Davis and Reuben Hersh, "Hilbert's Tenth Problem," in J. C. Abbot, ed., *The Chauvenet Papers*, vol. 2 (Washington: Mathematical Association of America, 1978), 554–71. (I owe this reference to Kenneth Manders.)

19. Gerald M. Weinberg in Paul Edwards, ed., *The Encyclopedia of Philosophy*, vol. 2 (New York: Macmillan and The Free Press, 1967), 173.

20. To be sure, there still remains the question: "Are there problems that people can solve *only* with the aid of computers?" But the emphatically affirmative answer that is virtually inevitable here involves no significant insult to human intelligence. After all, the same concession must be made with regard to reference aids of all sorts, the telephone directory being an example.

21. I am grateful to Gerald Massey, Laura Ruetsche, and especially Alexander Pruss for constructive comments and useful suggestions on a draft of this chapter.

22. To be sure, due care must be taken in construing the idea of greater performative capability. Thus, let C be the computer in question and let s be a generally computer-undecidable statement. Then consider the question:

(Q) Is it the case that at least one of the following two conditions holds? (i) You (i.e., the computer at work on the question Q now being posed) are C, (ii) s is true.

Since (i) obtains, C can answer this interrogation affirmatively. But since s is (by hypothesis) computer-undecidable, no other computer whatever can resolve Q. It might thus appear that no computer whatever could have greater general capability than another. However, its essential use of "you" prevents Q actually qualifying as "a question that C can answer but C' cannot." For in fact *different questions* are being posed—and different problems thus at issue—when the interrogation Q is addressed to C and to C'. For the example to work its intended damage we would need to replace (1) by one single fixed question that computer C would answer correctly with "yes" and any other computer C' would answer correctly with "no." And this is impossible.

Chapter 13

1. Roger A. MacGowan and Frederick I. Ordway 3d, *Intelligence in the Universe* (Englewood Cliffs, N.J.: Prentice Hall, 1966), 248.

2. "Apologie de Raimond Sebond," in *Les Essais de Michel de Montaigne*, ed. Pierre Villey, tome 2 (Paris: F. Alcan, 1922), 147.
3. Compare the discussion in Goesta Ehrensvärd, *Man on Another World* (Chicago and London: University of Chicago Press, 1965), 146–48.
4. His anthropological investigations pointed Benjamin Lee Whorf in much this same direction. He wrote: "The real question is: What do different languages do, not with artificially isolated objects, but with the flowing face of nature in its motion, color, and changing form; with clouds, beaches, and yonder flight of birds? For as goes our segmentation of the face of nature, so goes our physics of the cosmos" ("Language and Logic," in *Language, Thought, and Reality*, ed. J. B. Carroll [Cambridge: MIT Press, 1956], 240–41). Compare also the interesting discussion in Thomas Nagel, "What Is it Like to Be a Bat?" in *Mortal Questions* (Cambridge: Harvard University Press, 1976).
5. Thomas Kuhn, *The Structure of Scientific Revolutions* (Chicago: University of Chicago Press, 1962).
6. Georg Simmel, "Uber eine Beziehung der Kelektionslehre zur Erkenntnistheorie," *Archiv für systematische Philosophie und Soziologie*, vol. 1 (1895), 34–45 (see 40–41).
7. William James, *Pragmatism* (New York: Longmans, 1907).
8. See E. Purcell in *Interstellar Communication: A Collection of Reprints and Original Contributions*, ed. A. G. W. Cameron (New York and Amsterdam: W. A. Benjamin, 1963).
9. Paul Anderson, *Is There Life on Other Worlds?* (New York and London: Crowell-Collier Press, 1963), 130.
10. Christiaan Huygens, *Cosmotheoros: The Celestial Worlds Discovered—New Conjectures Concerning the Planetary Worlds, Their Inhabitants and Productions* (London: T. Childe, 1698), 41–43.
11. John A. Ball, "Extraterrestrial Intelligence: Where Is Everybody?" *American Scientist*, vol. 68 (1980), 565–663 (see 658).
12. Robert Nozick, "R.S.V.P.—A Story," *Commentary*, vol. 53 (1972), 66–68.
13. John A. Ball, "The Zoo Hypothesis," *Icarus*, vol. 19 (1973), 347–49.
14. See M. Calvin in A. G. W. Cameron, ed., *Interstellar Communication: A Collection of Reprints and Original Contributions* (New York and Amsterdam: W. A. Benjamin, 1963), 75.
15. George Gaylord Simpson, "The Nonprevalence of Humanoids," *Science* 143 (1964): 769–75; this article is chapter 12 of *This View of Life: The World of an Evolutionist* (New York: Harcourt, Brace, and World, 1964), see 773.
16. A. G. W. Cameron, ed., *Interstellar Communication: A Collection of Reprints and Original Contributions* (New York and Amsterdam: W. A. Benjamin, 1963).
17. Robert T. Rood and James S. Trefil, *Are We Alone? The Possibility of Extraterrestrial Civilization* (New York: Scribner, 1981).

18. Diels-Kranz 68 A 40 [for Leucippus and Democritus], *The Presocratic Philosophers,* trans. G. S. Kirk and J. E. Raven (Cambridge: Cambridge University Press, 1957), 411.

19. Alexandre Koyré, *From the Closed World to the Infinite Universe* (New York: Harper, 1958).

20. Christiaan Huygens, *Cosmotheoros: The Celestial Worlds Discovered—New Conjectures Concerning the Planetary Worlds, Their Inhabitants and Productions* (London: T. Childe, 1698), 76–77.

21. Roland Pucetti, *Persons: A Study of Possible Moral Agents in the Universe* (New York: Macmillan, 1969), cp. Ernan McMullin, "Forms of Life," *Icarus* 14 (1971), 291–94.

22. Loren Eiseley, *The Immense Journey* (New York: Random House, 1937).

23. Fred Hoyle, *The Black Cloud* (New York: Harper, 1957).

24. I. S. Shklovskii and Carl Sagan, *Intelligent Life in the Universe* (San Francisco: Holden-Day, 1966), 350.

Chapter 14

1. A collection of particularly fine articles on the subject is Harry Woolf, ed., *Quantification: A History of the Meaning of Measurement in the Natural and Social Sciences* (Indianapolis: Bobbs-Merrill, 1961). See also Brian Ellis, *Basic Concepts of Measurement* (Cambridge: Cambridge University Press, 1966).

2. See, for example, the introduction to David R. Kranz et al., *Foundations of Measurement,* vol. 1 (New York and London: Academic Press, 1971).

3. S. S. Stevens, "Measurement, Psychophysics, and Utility," in *Measurement: Definition and Theories,* ed. C. W. Churchman and P. Ratoosh (New York: Wiley, 1959), 18–63 (see 19).

4. M. R. Cohen and Ernest Nagel, *An Introduction to Logic and Scientific Method* (New York: Harcourt, Brace, and Company, 1934), 294. Compare Norman R. Campbell: "Measurement is the process of assigning numbers to represent qualities," *Foundations of Science* (New York: Dover, 1957).

5. One of the main reasons for being of the fashionable idea of a "system" is to play just exactly this role.

6. Herbert Dingle, "A Theory of Measurement," *British Journal for the Philosophy of Science,* vol. 1 (1950), 5–26.

7. In looking at the extent to which one quantity correlates with others one must do one's scorekeeping fairly. The number of noses in the cage correlates with the number of animals there. Well and good. But it does not *also* get credit for correlating with the number of tails, or ears, or hooves, etc.

8. For this distinction see N. R. Campbell, *Physics: The Elements* (Cambridge: Harvard University Press, 1920) and also *What Is Science?* (Cambridge: Cambridge University Press, 1921); reprint New York: Dover Publications, 1952.

9. Recent deliberations in the philosophy of science have done much toward showing that the attribution of observable features to the furnishings of nature is always theory-laden—part of a process in which our theoretical understanding of nature plays a significant role. But this is all the more so with respect to the attribution of measurable features, seeing that measurement is even more pronouncedly theory-involving than observation.
10. The issues are closely parallel because a measured feature of something can be looked upon as constituting a parametrized quality of it. As my Pittsburgh colleague Nuel Belnap has noted in discussion, all of the seven factors singled out in the previous quantity-oriented discussion have close analogues in the qualitative case in relation to natural kind versus random assemblage distinction. The difficulty of maintaining natural kindhood in the face of *logical* combinations (such as conjunction) exactly parallels the difficulty of maintaining actual measurementhood in the face of *mathematical* combinations (such as multiplication).
11. See Richard H. Shryock, "Quantification in Medical Science," in Harry Woolf, ed., *Quantification* (Indianapolis: University of Indiana Press, 1961). Note, incidentally, that "life expectancy" will not do the trick here, seeing that a person may well expect a long life on whose course he often or generally "feels miserable."
12. Testimony of Allan J. McDonald as reported in the *New York Times*, February 26, 1986, 16.
13. Sir William Thomson, "Electrical Units of Measurement," *Popular Lectures and Addresses*, 3 vols. (London: Macmillan, 1889–1891), vol. 1, 73.
14. See Peter Skrabanek and James McCormick, *Follies and Fallacies in Medicine* (New York: Prometheus Books, 1990), 27.
15. This chapter is an expanded revision of a discussion initially published in the author's *Satisfying Reason* (Dordrecht: Reidel, 1995), 71–83.

Chapter 15

1. Hilary Putnam's *Reason, Truth, and History* (Cambridge: Harvard University Press, 1981) contains interesting discussions of this issue.
2. F. C. S. Schiller, *Must Philosophers Disagree?* (London: Macmillan and Co., 1934), 5–7.
3. To be sure, the pursuit of the scientific world in some areas entails not only *price* but also *risks*. But only science itself can provide us with the information needed to calculate these in an intelligent way. Nothing said here conflicts with the idea that the pursuit of science has its costs, which must be weighed on the scale along with others.
4. See Martin Heidegger, *Being and Time*, ed. J. Macquarrie and E. Robinson (London: SCM Press, 1962), sect. 32.

5. Ludwig Wittgenstein, *Tractatus Logico-Philosophicus* (London: Routledge & Kegan Paul, 1922), sect. 6.52.
6. Lasek Kolakowski, *The Alienation of Reason*, trans. N. Guterman (Garden City, N.Y.: Doubleday, 1968), 204.
7. Paul Feyerabend, *Against Method* (London: Humanities Press, 1975). See also his "How to Defend Society Against Science," *Radical Philosophy*, no. 11 (1975), 84–129.
8. G. W. F. Hegel, *Phaenomenologie des Geistes*, ed. J. Hoffmeister (Hamburg, 1952), 63–66. A. V. Miller translation (Oxford: Oxford University Press, 1977), sect. 73–76.

INDEX

Adams, Henry Brooks, 63, 260n33
Allen, Thomas Barton, 219
Allison, D. C. S., 265n2
Amaldi, Edoardo, 258n14, 268n1, 271n3
Anderson, Paul, 219, 275n9
Aristotle, 9, 20, 41, 52, 67, 97, 136–38
Asimov, Isaac, 220

Bacon, Francis, 39, 105
Baer, K. E. von, 44
Ball, John A., 219, 275n11, 275n13
Barrow, John D., 269n9
Beck, Lewis White, 220
Belnap, Nuel, 277n10
Bentley, Richard, 69
Berrill, N. J., 220
Berstein, Jeremy, 265n1
Bohm, David, 78-79, 262n25, 266n11
Bohr, Neils, 32, 255n11
Boltzmann, Ludwig, 52, 73
Bracewell, Ronald, N., 220
Bradley, F. H., 156
Bragg, W. H., 255n11
Bragg, W. L., 255n11
Brandon, Robert N., 256n2
Brennerman, Hans, J., 273n5
Breuer, Reinhard., 220
Breuer, Thomas, 269n10
Bridgman, P. W., 224
Bromley, D. A., 263n36, 264n36, 271n4
Bruno, Giordano, 215

Campbell, Norman R., 276n4, 276n8
Capek, Milic, 255n6
Cassirer, Ernst, 267n5
Casti, John, L., 265n1

Champlain, T. S., 269n10
Church, Alonzo, 180, 187, 273n11
Cicero, 106, 203, 266n13
Cohen, M. R., 276n4
Collingwood, R. G., 106, 253n5, 266n14
Comte, Auguste, 124, 175
Cotes, Roger, 69, 98
Curry, Charles Emerson, 255n9

Dalton, John, 20, 32, 73
Darwin, 10, 98, 122
Davis, Martin D., 272n2, 274n18
Descartes, René, 20
deSolla Price, Derek J., 272n7
Dick, Steven J., 220
Dilthey, Wilhelm, 109
Dingle, Herbert, 224, 276n6
Dirac, Paul, 255n11
Disraeli, Benjamin, 236
Dole, Stephen H., 220
Donne, John, 215
Drachmann, A. G., 174
Drake, Frank D., 220
Dreyfus, Hubert, L., 189
Dupré, John, 258n13

Edwards, Jonathan, 269n7
Ehrensvaerd, Goesta, 220, 275n3
Einstein, Albert, 69, 95, 215, 255n11
Eiseley, Loren, 217, 276n22
Ellis, Brian, 276n1
Ewell, Raymond, 55, 258n19

Fermi, Enrico, 255n11
Feyerabend, Paul, 71–72, 249, 254n3, 256n7, 261n14, 278n7

Feynman, Richard, 25, 255n12, 262n23, 270n5
Fiebig, Hans, 256n10
Firsoff, V. A., 220
Foster, Sir Michael, 261n17
Franklin, Benjamin, 73

Gale, George, 268n18
Gale, Richard, M., 268n6
Galen, 72
Galileo, 9, 33, 41, 115, 125
Gallie, W. B., 265n4
Gavvay, Allen., 220
George, William, 258n21
Gilfillan, S. Colum, 266n6
Glass, Bentley, 78, 262n21
Gödel, Kurt, 180, 187, 191, 265n3, 272n4, 273n10, 274n18
Goethe, J. W. von, 239, 263n31
Grünbaum, Adolf, 253n4, 254n2

Haeckel, Ernest, 111, 114–15, 267n6, 267n7
Handy, Rollo, 267n6
Harré, Rom, 271n11
Hart, M. H. , 220
Hartley, David, 68
Hawking, S. W., 262n23., 270n4, 270n5
Hegel, G. W. F., 150, 156, 246, 249., 278n8
Heidegger, Martin, 246, 277n4
Heisenberg, 103, 255n11
Hempel, Carl G., 117–18, 135, 267n9, 268n4, 268n5
Herrmann, Joachim, 220
Herschel, John, 85, 263n33
Hersh, Reuben, 274n18
Hertz, Heinrich, 99
Hoare, C. A. R., 265n2
Hobbes, Thomas, 39, 256n10
Hoerner, Sebastian von, 220
Hofstadter, Douglas, 272n2
Holland, John H., 256n2
Holton, Gerald, 272n6
Hoyle, Fred, 218, 221, 276n23
Huang, Su-Shu, 220
Hume, David, 118, 269n7
Husserl, Edmund, 148
Huygens, Christiaan, 73, 204, 217, 221, 275n10, 276n20

Inhelder, Bärbel, 254n7

James, William, 202, 254n15, 275n7
Jankey, Karl, 95
Jeans, Sir James, 221
Jeffrey, Eber, 270n6
Jevons, W. Stanley, 12, 14, 22, 83, 86, 122, 175, 254n10, 254n12, 255n5, 262n26, 263n35, 264n41, 268n11, 272n11
Jolly, Philip von, 24

Kant, Immanuel, 5, 10, 12–16, 18, 66, 97, 112, 124, 130, 149, 159, 216, 219, 250, 253n5, 254n11, 257n9, 271n8
Kaufmann, Stuart, 257n3
Kelvin, Lord 7, 23, 73, 238
Keynes, J. M., 71
Kitcher, Philip, 270n15
Kolakowski, Lasek, 278n6
Korach, M., 272n9
Koyré, Alexandre, 215, 276n19
Kranz, David R., 276n2
Kuhn, Thomas, 38, 71–72, 201, 256n9, 261n14, 266n10, 275n5

Lagrange, J. L. de, 83
Lakatos, Imre, 72
Lambert, J. H., 105
Laplace, P. S. de, 83
Laudan, Larry, 21, 105, 255n4, 255n14, 255n1, 260n3, 261n11, 261n16, 262n20, 266n12
Laue, Max von, 255n11
Le Sage, Georges, 68, 105, 260n6
Legendre, A. M., 124
Lehrer, Keith, 271n13
Leiber, Theodore, 273n6
Leibniz, G. W., 12, 118, 130, 267n12
Lem, S., 221
Leslie, John, 267n14, 268n18
Levin, Michael, E., 256n3
Lister, Joseph, 72
Locke, John, 105

MacGowan, Roger A., 219, 221, 274n1
Machlup, Fritz, 258n22
Manders, Kenneth, 274n18
Margenau, Henry, 271n3
Massey, Gerald, 274n21

INDEX

Maxwell, James Clerk, 99
McCabe, J., 267n6
McCormick, James, 277n14
McDonald, Allan J., 277n12
McMullin, Ernan, 75, 221, 261n19
Menard, H. W., 258n20
Mendenhall, T. C., 23, 255n7
Michelson, 255n11
Mill, John Stuart, 175
Millikan, Robert, 69, 95, 260n9
Montaigne, M. E. de, 198

Nagel, Ernest, 275n4
Nagel, Thomas, 276n4
Newton, Isaac, 9, 12, 20, 33, 52–53, 69, 73, 97–98, 105, 133, 154, 199
Niiniluoto, Ilkka, 260n2, 264n38
Norris, Keith, 272n8
Nozick, Robert, 221, 268n15, 275n12

Oppenheim, Paul, 135, 268n4, 268n5
Ordway, Frederick I., III., 219, 221, 274n1
Ostwald, Wilhelm, 255n9

Paracelsus, 72
Pascal, Blaise de, 215, 262n29
Passmore, John, 253n2
Pasteur, Louis, 72, 96, 98
Pauli, Wolfgang, 255n11
Pearson, Karl, 115
Peirce, Charles Sanders, 54, 67–68, 80, 86, 110, 124, 155, 159, 167, 206, 257n9, 259n29, 260n4, 260n1, 260n2, 268n19
Penrose, Roger, 273n15
Penzias, Arno A., 95
Petley, B. W., 257n11
Piaget, Jean, 254n7
Planck, Max, 24, 52, 255n8, 255n11, 259n29
Plato, 142–143, 156, 206
Poincaré, Henri, 47
Popper, Karl A., 20–21, 72, 136, 254n1, 255n1, 261n15, 265n3, 265n4, 269n11
Pour-El, N. B., 272n2
Powell, Baden, 109, 125, 266n8, 266n16, 268n20
Price, Derek J., 256n5, 258n16, 258n17
Priestley, Joseph, 68, 260n5
Provost, Pierre, 260n6
Pruss, Alexander, 268n6, 274n21
Ptolemy, 33, 52

Pucetti, Roland, 217, 221, 276n21
Purcell, Edward, 275n8
Putnam, Hilary, 277n1

Quay, Paul M., 256n9
Quine, W. V. O., 72, 261n13, 261n14, 264n42

Rapp, Friedrich, 256n9
Rayleigh, J. W. S., 7
Reichenbach, Hans, 257n6
Reymond, Emil du Bois, 106–07, 111, 113–15, 267n5, 268n17
Richards, J. I., 272n2
Rogers, Hartley, 273n11
Rood, Robert T., 221, 275n17
Rosenberg, Alex, 266n4
Rowe, William, L., 267n13, 268n6
Royce, Josiah, 156
Ruetsche, Laura, 274n21
Rutherford, Ernest, 24, 32

Sagan, Carl, 221, 276n24
Salmon, Wesley, 268n4
Sarton, George, 260n10
Schiller, F. C. S., 243–44, 277n2
Schrödinger, Erwin, 255n11
Sellars, Wilfred, 271n11
Shapley, Harlow, 221
Shklovskii, I. S., 221, 276n24
Shryock, Richard H., 277n11
Simmel, Georg, 202, 275n6
Simon, Herbert A., 263n30
Simpson, George Gaylord, 213, 222, 275n15
Skrabanek, Peter, 277n14
Soddy, Frederick, 95, 265n2
Spencer, Herbert, 44, 256n1
Spinoza, Benedictus de, 156, 262n24
Stent, Gunther S., 25, 255n13, 262n23, 270n5
Stevens, S. S., 224, 276n3
Sullivan, Walter, 222

Thomson, J. J., 24, 255n10, 255n11
Tipler, Frank T., 269n9
Trefil, James S., 221, 275n27
Turing, Alan, 89, 180, 188
Tyndall, John, 86

Ulam, Stanislaw M., 259n28
Urbach, Peter, 265n3, 266n4

Vaizey, John, 272n8
Vigier, Jean Paul, 80, 262n29, 263n29
Visher, S. S., 258n15
Von Neumann, John, 103, 194

Wegener, Alfred, 70
Weinberg, Alvin, M., 55
Weinberg, Gerald M., 274n19
Weinberg, Steven, 128, 258n14, 259n23, 268n1
Weizsäcker, C. F. von, 258n14, 268n1
Wells, H. G., 210
Whewell, Williiam, 261n10

Whitehead, A. N., 102, 266n9
Whorf, Benjamin Lee, 275n4
Wigner, E. P., 84, 262n27, 263n29, 264n37, 264n39, 271n3
Wilkinson, Sir Denys, 262n28, 271n1
Wittgensein, Ludwig, 247, 278n5
Wollaston, William, 67

Yukawa, Hideki, 25, 255n11

Ziman, John N., 261n12, 266n9